Sustainable Agriculture Reviews

Volume 37

Series Editor
Eric Lichtfouse
Aix-Marseille Université, CNRS, IRD, INRA
Coll France, CEREGE
Aix-en-Provence, France

Other Publications by Dr. Eric Lichtfouse

Books
Scientific Writing for Impact Factor Journals
https://www.novapublishers.com/catalog/product_info.php?products_id=42242

Environmental Chemistry
http://www.springer.com/978-3-540-22860-8

Sustainable Agriculture
Volume 1: http://www.springer.com/978-90-481-2665-1
Volume 2: http://www.springer.com/978-94-007-0393-3

Book series
Environmental Chemistry for a Sustainable World
http://www.springer.com/series/11480

Sustainable Agriculture Reviews
http://www.springer.com/series/8380

Journal
Environmental Chemistry Letters
http://www.springer.com/10311

Sustainable agriculture is a rapidly growing field aiming at producing food and energy in a sustainable way for humans and their children. Sustainable agriculture is a discipline that addresses current issues such as climate change, increasing food and fuel prices, poor-nation starvation, rich-nation obesity, water pollution, soil erosion, fertility loss, pest control, and biodiversity depletion.

Novel, environmentally-friendly solutions are proposed based on integrated knowledge from sciences as diverse as agronomy, soil science, molecular biology, chemistry, toxicology, ecology, economy, and social sciences. Indeed, sustainable agriculture decipher mechanisms of processes that occur from the molecular level to the farming system to the global level at time scales ranging from seconds to centuries. For that, scientists use the system approach that involves studying components and interactions of a whole system to address scientific, economic and social issues. In that respect, sustainable agriculture is not a classical, narrow science. Instead of solving problems using the classical painkiller approach that treats only negative impacts, sustainable agriculture treats problem sources.

Because most actual society issues are now intertwined, global, and fast-developing, sustainable agriculture will bring solutions to build a safer world. This book series gathers review articles that analyze current agricultural issues and knowledge, then propose alternative solutions. It will therefore help all scientists, decision-makers, professors, farmers and politicians who wish to build a safe agriculture, energy and food system for future generations.

More information about this series at http://www.springer.com/series/8380

Inamuddin • Abdullah M. Asiri • Eric Lichtfouse
Editors

Sustainable Agriculture Reviews 37

Carbon Sequestration Vol. 1 Introduction and Biochemical Methods

Springer

Editors

Inamuddin
Chemistry Department, Faculty of Science
King Abdulaziz University
Jeddah, Saudi Arabia

Department of Applied Chemistry,
Faculty of Engineering and Technology
Aligarh Muslim University
Aligarh, India

Abdullah M. Asiri
Chemistry Department, Faculty of Science
King Abdulaziz University
Jeddah, Saudi Arabia

Eric Lichtfouse
Aix-Marseille Université
CNRS, IRD, INRA, Coll France,
CEREGE
Aix-en-Provence, France

ISSN 2210-4410 ISSN 2210-4429 (electronic)
Sustainable Agriculture Reviews
ISBN 978-3-030-29297-3 ISBN 978-3-030-29298-0 (eBook)
https://doi.org/10.1007/978-3-030-29298-0

This Springer imprint is published by the registered company Springer Nature Switzerland AG.
The registered company address is: Gewerbestrasse 11, 6330 Cham, Switzerland

Preface

Carbon dioxide flow out into the environment as an outcome of substantial dependence on nonrenewable sources of energy and is tremendously increasing environmental pollution. In the current scenario, carbon dioxide sequestration is a key solution to overcome the issues of global warming, which can extraordinarily diminish greenhouse carbon dioxide discharge from power plants and industries. There are various methods available for the sequestration of carbon dioxide, including biological, chemical, and physical processes. However, there is an exigent demand to improve available methods of carbon dioxide sequestration. The advancements being made in this technology are incredibly impacted by the rise of novel methods and materials.

Carbon Sequestration Vol 1: Introduction and Biochemical Methods is compiled, keeping the goal to provide information regarding carbon dioxide emission sources and environmental issues. It is also devoted to introducing the methods used for carbon dioxide utilization, storage, analysis, modeling, and optimization . Some of the biochemical methods of carbon dioxide sequestration, including forestry, biomineralization , geo-chemo-mechanical, mangrove plantation, and biowaste, are discussed in details. It is a unique reference book, incredibly well-organized, and a fundamental resource and reference guide for environmental geoscientists, engineers, industrial experts, faculty, R&D professionals, postgraduate and undergraduate students, environmental scientists, and researchers. Based on thematic topics, the book edition contains the following eight chapters:

Chapter 1 gives a deeper insight into the major components of the carbon capture and storage technology. The mechanisms, challenges, and issues of the carbon capture and storage technology are also discussed.

Chapter 2 tries to lay down the various sources of CO_2 emissions and environmental issues arising from CO_2 emissions.

Chapter 3 reviews systems for CO_2 capture and compression and options for CO_2 utilization and storage. The design, optimization, and utilization of carbon capture storage supply chain are also discussed.

Chapter 4 reviews the role of the terrestrial biosphere, mainly the forests in anthropogenic carbon sequestration. It examines the influence of the environment and climate variables of the global carbon cycle of the terrestrial ecosystems. It also assimilates the knowledge in global carbon budget generated in the past few decades.

Chapter 5 summarizes the processes and mechanisms of microbially mediated and plant-mediated biomineralization and highlights potential applications of biomineralization in CO_2 sequestration with summery of key knowledge gaps and future research directions.

Chapter 6 reviews the current state of knowledge regarding CO_2 and shale interactions, their potential impacts on shale properties, and groundwater quality. The characterization of shale gas and CO_2 is also summarized. The major interaction mechanisms between CO_2 and shale and their impact on rock properties and groundwater quality are surveyed. Finally, the open questions in the field are emphasized, and new research needs are highlighted.

Chapter 7 reviews the plantation methods and restoration techniques for enhanced blue carbon sequestration by mangroves.

Chapter 8 summarizes the use of biowastes in the sequestration of carbon evolved from environmental compartments. It provides a brief overview of sources and environmental impacts of biowastes. Finally, the use of biowaste for carbon sequestration is discussed.

Jeddah, Saudi Arabia — Inamuddin
Jeddah, Saudi Arabia — Abdullah M. Asiri
Aix-en-Provence, France — Eric Lichtfouse

Contents

Chapter 1
Introduction to Carbon Dioxide Capture and Storage

Arshad Raza and Raoof Gholami

Abstract CO_2 storage in deep geological formations such as depleted hydrocarbon reservoirs, saline aquifers, and methane coal beds has been recognized as an effective method to prevent climate change in the near future. This chapter gives a deeper insight into the major components of the carbon capture and storage technology as a major greenhouse gas mitigation approach implemented worldwide. The mechanisms, challenges, and issues understood so far linked to this technology are discussed. It appears that if a storage project can be safely implemented given the precautions mentioned in this chapter, a great step can be taken towards a better future for the next generation.

Keyword Greenhouse gases · Carbon dioxide · Carbon capture and storage technique · Injectivity · Containment · Monitoring

1.1 Introduction

Greenhouse gases have been increasingly released into the atmosphere for the past decades, causing a significant increase in earth temperature and ultimate climate change (Arocho et al. 2014). Extreme emissions of these gases induced by burning of the fossil fuels in the power plants, refineries, petrochemical industry, etc. may damage agricultural crops, forest species and ecosystem (Afroz et al. 2003; Lin and Ahmad 2017; Saxena and Bhargava 2017). Greenhouse gases often contain 76% CO_2 which has a significant impact on the environment and ecosystem compared to other gases (Raza et al. 2018, 2019). Technically, CO_2 is released from power plants (around 40%), transportation sector (about 20%) as well as construction and farming sectors (around 17%) (Anwar et al. 2018). Currently, the concentration of CO_2 in the

A. Raza
Department of Petroleum Engineering, University of Engineering and Technology (UET), Lahore, Pakistan

R. Gholami (✉)
Department of Petroleum Engineering, Curtin University, Sarawak, Malaysia
e-mail: Raoof.Gholami@Curtin.edu.my

Inamuddin et al. (eds.), *Sustainable Agriculture Reviews 37*, Sustainable Agriculture Reviews 37, https://doi.org/10.1007/978-3-030-29298-0_1

atmosphere has reached 410 parts per million which are almost 20% more than their standard threshold (Wennersten et al. 2015).

There have been many approaches presented so far to mitigate the emission of greenhouse gases into the atmosphere which includes (1) burning fossil fuels with low carbon content, (2) employing low carbon energy sources, (3) sequestering CO_2 through natural sinks and (4) storing CO_2 in deep geological formations (IPCC 2005). However, the storage of CO_2 using the carbon capture and storage technology has been recognized as the most effective mitigation approach to reduce the amount of CO_2 released in the atmosphere. In this method, CO_2 is captured from the power plants, transported through the pipelines and stored into suitable geological sites such as depleted hydrocarbon reservoirs, saline aquifers, and coal beds. To store a large volume of CO_2 for the reduction of CO_2 emissions, however, a comprehensive assessment of the capturing techniques and available geological formations is required (IPCC 2007) although some challenges and economic issues would ultimately emerge (Dindi et al. 2019). As a matter of fact, there are many concerns related to the success of the carbon capture and storage technology such as the selection of the best capturing and separation methods, the proper characterization of the storage sites, the structural integrity of the seals, etc. which have not been totally understood yet. This chapter attempts to provide a deeper insight into the concept of the carbon capture and storage technology and the mechanisms involved which may need further studies.

1.2 Carbon Dioxide

CO_2 is a major greenhouse gas produced from Coal (43%), oil (37%) and natural gas (20%). It can appear in solid, liquid, gas and supercritical states depending on the pressure and temperature of the medium (See Fig. 1.1). Given the fact that the storage sites are often deeper than 800 m, the temperature and pressure (i.e. 31.10 °C and 1070 psi) push CO_2 to appear as a supercritical fluid. Supercritical CO_2 has the diffusivity of the gas but the density of the liquid and can occupy lesser space compared to its other phases. It can also dissolve in the formation brine and produce carbonic acid which decreases the pH of the solution in the reservoir and pose integrity issues.

CO_2 is mainly produced from the power generation and heat supply sectors with a total amount of 10.5 gigatonnes in 2004 which may get close to 14.6 gigatonnes in 2030. None energy sectors are also playing a significant role in the emission of CO_2 which may contribute to global warming by producing 10.6 gigatonnes CO_2 in 2030. This indicates the fact that some technological approaches must be employed to mitigate the emission of CO_2 and other greenhouse gases. Carbon Capture and Storage technology is one of these effective methodologies which is presented in the next section.

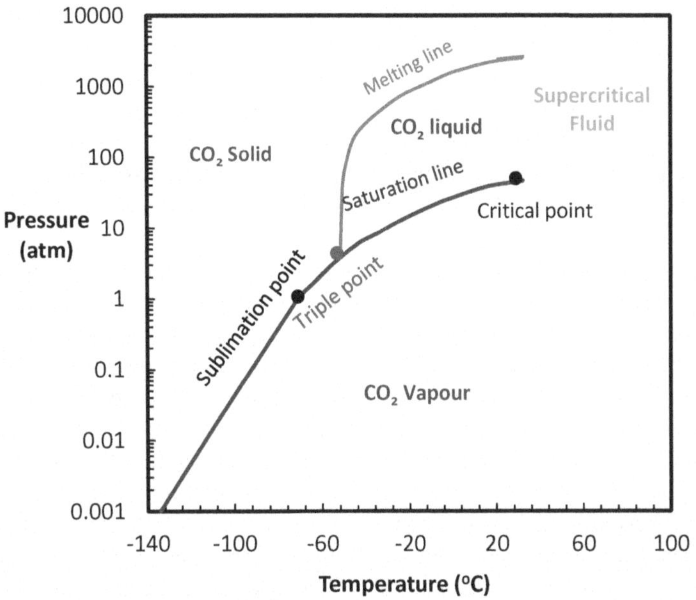

Fig. 1.1 Pressure-Temperature phase diagram of carbon dioxide showing different phases of CO_2 under various pressure and temperature conditions

1.3 Carbon Dioxide Capture and Storage Technology

Carbon Dioxide Capture and Storage technology consists of four major steps where CO_2 is separated/captured from industrial sites, transported to a suitable site, injected/stored into proper geological formations and monitored for safety purposes. In this section, these four stages are presented and discussed in details.

1.3.1 Capturing and Separation

It is essential to evaluate the combustion process, absorption and separation of CO_2 prior to initiating a carbon capture and storage project (Leung et al. 2014). Capturing CO_2 from large emission sources such as fossil fuel and processing plants of iron, steel, and cement is the first step in this process. It should be reminded that capturing from a small emission source would be more difficult and expensive. Typically, three major capturing technologies are used for the capturing which include post-combustion, pre-combustion and oxy-fuelling methods (Bennaceur et al. 2008; IPCC 2005; Leung et al. 2014; Moazzem et al. 2012; Songolzadeh et al. 2014). The post-combustion is the most effective approach where CO_2 is captured from the combustion process by a chemical solvent but this method cannot be used for capturing a large volume of CO_2 (Anwar et al. 2018). In the pre-combustion method, hydrogen-rich fuel is pre-treated in a gasifier under the high pressure of oxygen. Considering the fact that fuel gas often has a huge concentration of CO_2, the separation of this capturing method is easier than others (IPCC 2005). In the oxyfuel method, on the other hand, pure oxygen is employed for capturing method and air is no longer considered. This capturing method is technically an efficient technology for CO_2 removal but due to the high concentration of sulfur dioxide in the flue gas, it is recognized as an energy intensive and corrosive method (Araújo and de Medeiros 2017).

There are also approaches proposed to capture CO_2 from flue gas in the post-combustion process which are referred to as the separation methods. These methods mainly consist of absorption, adsorption, membrane separation, hydrate-based separation, cryogenic distillation, and chemical looping. Absorption, for instance, is done by flowing the flue gas through the solvent to get CO_2. Zeolite is used in the physical adsorption as a catalyst for separating CO_2 in the power plants and natural gas treatment systems (Choi et al. 2009). The carbon-based adsorbing method is done on coal where activated carbon and charcoal are obtained under high-pressure. Membrane separation is done by ceramics, the metallic or polymeric membrane with the efficiency of 90% for separating CO_2 (Leung et al. 2014). Hydrate-based separation is a new and low-cost approach used to remove CO_2 from the exhaust gas by generating hydrates at high-pressure conditions. Cryogenic distillation works based on the boiling point of CO_2 and separates it at the temperature of −100 °C to −135 °C. Chemical looping combustion uses metal oxides for the capturing process where metal and oxidizes are produced but recycled at the end (Anwar et al. 2018). Table 1.1 summarises these capturing and separation technologies.

It should be noted that there are risks involved in the CO_2 capturing systems given the fact that flue and vent gases are emitted to the atmosphere or liquid wastes are produced. As such, rules and regulations of each region chosen for the carbon capture and storage project must be understood before initiating any capturing processes. The capture costs such as plant design, operation, and financial factors are the other important parameters to consider when it comes to choosing a capturing methodology.

1.3.2 Transport

Upon capturing, CO_2 is transported to suitable geological sites for storage. The volume of CO_2 after compression dictates the transportation method which can be done by trucks, ships or pipelines (Surface or Underwater) (Leung et al. 2014). Some of the successful transportation projects done through the pipeline are canyon Reef, Bravo Dome, Cortez, Sheep mountain and Weyburn. It should be reminded that dry supercritical CO_2 cannot corrode the steel commonly used to manufacture the pipelines for transportation as long as the humidity is less than 60%. However, impurities (e.g., nitrogen oxides, sulfur dioxide, and hydrogen sulfide) in the presence of water in the tank may lead to corrosion and hydrate formation. They may

Table 1.1 Summary of capturing and separation techniques used in the carbon capture and storage technology (Cuéllar-Franca and Azapagic 2015)

Capture option	Separation technology	Method	Applications
Pre-conversion	Absorption by physical solvent	Selexol, rectisol	Power plants
	Absorption by chemical solvents	Amine-based solvent, e.g. monoethanolamine (MEA)	Ammonia production
	Adsorption by porous organic frameworks	Porous organic frameworks membranes	Gas separations
Post-conversion	Absorption by chemical solvents	Amine-based solvent, e.g. monoethanolamine (MEA), diethanolamine (DEA), and hindered amine (KS-1)	Power plants; iron and steel industry; cement industry; oil refineries
		Alkaline solvents, e.g. Sodium hydroxide (NaOH) and ca $(OH)_2$	
		Ionic liquids	
	Adsorption by solid sorbents	Amine-based solid sorbents	No application reported
		Alkali earth metal-based solid sorbents, e.g. calcium carbonate ($CaCO_3$)	
		Alkali metal carbonate solid sorbents, e.g. sodium carbonate (Na_2CO_3) and Potassium carbonate (K_2CO_3)	
		Porous organic frameworks – Polymers	Power plants
	Membrane separation	Polymeric membranes, e.g. polymeric gas permeation membranes	Power plants; natural gas sweetening
		Inorganic membranes, e.g. zeolites	
		Hybrid membranes	
	Cryogenic separation	Cryogenic separation	Power plants
	Pressure/vacuum swing adsorption	Zeolites	Power plants; iron and steel industry
		Activated carbon	
Oxy-fuel combustion	Separation of oxygen from air	Oxy-fuel process	Power plants; iron and steel industry; cement industry
		Chemical looping combustion	Power plants
		Chemical looping reforming	Power plants; syngas production and upgrading

also change the pressure during transportation and cause unintended multiphase flow and unanticipated corrosion. Thus, the thermodynamic behavior of CO_2 with impurities from "typical" to transient condition must be well understood before the deployment of any carbon capture and storage projects (Jensen et al. 2014).

1.3.3 Injection and Storage

Before injection, a suitable site must be chosen and assessed at basin and reservoir scales to ensure that CO_2 can be remained confined. In fact, these storage sites must be chosen wisely as they should hold the injected CO_2 for thousands of years without any leakage to surface or subsurface resources. Suitable geological formations are deep saline aquifers, oil and gas reservoirs, and deep coal beds (Raza 2017). Although deep oceans have a large storage capacity, they often suffer from technical, safety, liability, and environmental issues (Leung et al. 2014). Coal beds have also unknown storage integrity, technical feasibility and relative capacity which may need further assessment before being recognized as a good storage site. Aquifer, on the other hand, are one of the attractive options due to the high storage capacity, but a detailed assessment of their integrity, cost, storage, and technical feasibility must be done before making any decisions (Herzog et al. 1997). There have also been some discussions around CO_2 injection in basalts or organic-rich shale considering their mineral trapping potential and adsorption capacity, but the low permeability of these reservoirs makes them an unlikely choice (White and McGrail 2003). As such, depleted hydrocarbon reservoirs are the best choice as they have held hydrocarbon for a long period of time and are probably able to do the same for CO_2 once injected. They also have the infrastructure required to inject CO_2. Technically speaking, a suitable storage site must have: i) sufficient capacity for storage, ii) good injectivity for injection with the rate required and iii) strong containment to avoid leakage/integrity issues. To have these three requirements, the formation chosen for the storage must be deep enough (>800 meters) with a thick seal rock to keep CO_2 at the supercritical state without initiation/reactivation of preexistence fractures/faults (Orr Jr 2009).

As it was mentioned earlier, storage site must be screened at different scales (i.e., basin, regional, and reservoir scales) to confirm their suitability for injection. At the reservoir scale, key aspects of the storage are evaluated by analytical, numerical and experimental approaches (Raza 2017). Key CO_2 storage aspects evaluated at this stage are storage capacity, injectivity, trapping mechanisms, and containment which are described as below:

- Storage capacity is the total usable storage volume of the site and cannot be easily determined in aquifers since different mechanisms are simultaneously involved (Bachu et al. 2007). Surface wettability of the reservoir and interfacial tension between CO_2 and water are the most important parameters which must be evaluated when the capacity of storage sites is estimated (Iglauer et al. 2015).

- CO_2 injection at a feasible injection rate without pressurizing the containment system is referred to as injectivity. It is controlled by the porosity, permeability, thickness, and heterogeneity of the formations (Han et al. 2011). During and after injection, low permeable seals act as a trap for the injected fluid. It should be noted that CO_2 migrates upward towards the seal as an immiscible phase considering the buoyancy and viscosity effects. Thus, as CO_2 flows to the top of the storage site, it is capillary trapped by the microscopic forces. Reservoir pressurization due to the injection initiates dissolution trapping for the bottom part of the plume when CO_2 dissolves into the brine (IPCC 2007). Dissolved CO_2 generates a carbonic acid along with bicarbonate ions which chemically precipitates rock minerals over time (Bachu et al. 1994). During the first few decades, structural and residual trappings take place while the dissolution and mineral trapping continue for centuries (Irfan et al. 2018). In coals, however, adsorption is the major trapping mechanism and supports a large capacity of injection (Zhao et al. 2016). However, technically speaking, capillary trapping is a key mechanism for retaining CO_2 in the reservoir in a short period of time before other trapping mechanisms can come to the play (Raza 2017). Several parameters, however, control the capillary trapping including permeability, thermodynamic properties of CO_2-H_2O and heterogeneity which must be understood before choosing the storage sites (Raza 2017).
- Containment as the last key storage aspect ensures the safety of injected CO_2 in the long term (IPCC 2007). It should be reminded that upon injection, some geomechanical issues might be induced due to the pressure buildup such as changes of the stress state, compaction, fault reactivations, and fractures generation (Raza et al. 2016). Proper selection of well position and production rate may help to control the pressure and prevent these geomechanical issues (Chadwick et al. 2009). CO_2 may also dissolve in formation water and interact with the caprock which can fail the caprock and/or reactivate the fracture/fault systems in the long term (Varre et al. 2015). These irreversible mechanical changes may create leakage pathways for CO_2. Thus, a proper assessment of storage sites for geomechanical integrity issues caused by the pressure build-up or geochemical interactions must be done on the field scale.

It should be reminded that like the capturing stage, the presence of impurities in the CO_2 stream can cause issues during storage, especially for the capacity estimation. It also has impacts on the trapping (dissolution and precipitation) mechanisms of CO_2 and its compressibility which in turn reduces the amount of CO_2 that could be stored in a given site.

1.3.3.1 Integrity Issues

There are several major integrity issues that may take place during and after CO_2 injection. One of the major issues reported so far is the degradation of class G (Portland) cement commonly used to plug the injection intervals. It is induced by the

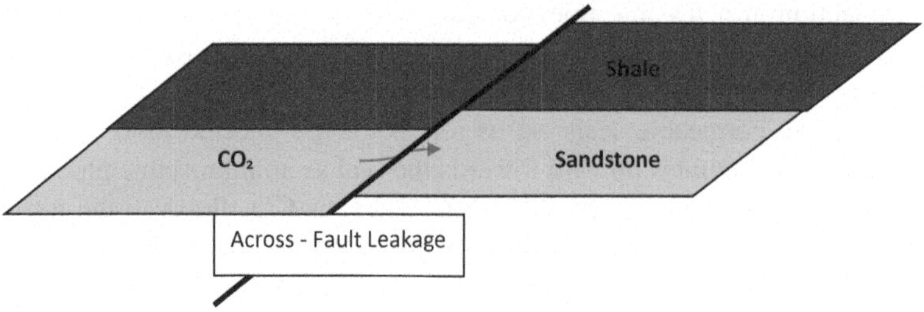

Fig. 1.2 Across – fault leakage from CO_2 storage sites where the pressure accumulation in the fault zone can cause fault reactivations upon leakage

geochemical interaction of the cement with supercritical CO_2 and carbonic acid in the presence of brine which can create a major leakage pathway for seepage of CO_2 from the storage site (Abid et al. 2018).

Another major integrity issue is the across fault leakage from the storage sites that may take place when the buoyancy pressure of the injected CO_2 becomes higher than the capillary pressure of the rocks surrounding the faults in the storage site. Faults are often considered as impermeable seals but they are membrane seal on many occasions. Surface wettability and pore throat of the rocks formed around a fault play an important role under these circumstances and must be accurately determined. Although the pressure accumulation in the fault zone is not often high enough to cause fault reactivations, the leakage induced can contaminate subsurface resources (See Fig. 1.2). However, there are scenarios where the two-phase permeability of a fault and the caprock is low, resulting in long time scales until CO_2 can breakthrough. The pressure building up along the fault surface, on these occasions, may significantly reduce the shear strength (due to the reduction of the normal stress) and cause reactivation (Rutqvist et al. 2007). This reactivation may lead to migration of CO_2 into other formations, environmental contamination, and even loss of the storage site.

1.3.4 Monitoring

To monitor the migration of injected CO_2, tools and techniques developed for the exploration and production of hydrocarbon reservoirs are often applied. This is done to (1) ensure that the injection well is capable of handling the fluid injected without having any leakage issues (2) determine the amount of CO_2 stored by different mechanisms, (3) make sure that the storage capacity is properly used, (4) verify that the fluid injected is remained confined and (5) detect leakages in the system and provide mitigation approaches.

Approaches developed for the monitoring purpose are generally divided into two categories of (1) monitoring the injection rate and pressure and (2) monitoring CO_2 distribution in subsurface layers. In the first technique, gauges or orifice meters are considered to determine bottom hole and wellhead pressure while seismic methods such as time-lapse 3D reflection, passive method and cross-well seismic imaging are known as the best techniques for monitoring the CO_2 plume migration. Many other methods such as applying tracers, sampling the reservoir brines and soil gas sampling can also be used to monitor CO_2 storage sites upon injection (Herzog 2009).

1.4 Technological and Scientific Concerns

Although there has been a significant development in the carbon capture and storage projects implemented worldwide, there are still several gaps in knowledge linked to this technology that must be resolved. Some of the gaps include:

- Although different capturing techniques have been developed and successfully employed, a better and cheaper approach is still in demand. Project management and integration of capturing, transport and storage do not seem to be very well understood on many occasions. This large-scale knowledge and experience are required to have a better estimation of the project cost at the early stage of implementation.
- More studies on the optimum distance between CO_2 capturing source to a suitable storage site are required for better cost management.
- Estimation of the capacity of storage sites is still a subject of interest and needs more studies for a better injection, migration and monitoring purposes.
- Ocean has been suggested as a potential place for storage but limited knowledge has been gained so far about the ecological impact of such action. Required methods and techniques for detecting CO_2 plum and its monitoring in the marine environment are also not in place yet.
- Roles and regulations for CO_2 storage have not been very well established in many countries. Planning and implementation of a carbon capture and storage project in these countries may take longer than expected.

1.5 Summary

Excessive concentration of greenhouse gases, particularly CO_2, in the atmosphere is raising the alarm of global warming and climate changes. Carbon capture and storage technology is an efficient way to store a large volume of CO_2 in a subsurface geologic formation and stabilize the earth's temperature. In this chapter, attempts were made to look into the components of the carbon capture and storage technology. Although there have been successes reported to the CO_2 projects implemented worldwide, there are still many concerns and gap in knowledge when it comes to capturing, transporting, storing and monitoring CO_2 in subsurface layers. It seems that more studies are required to ensure that CO_2 can be cheaply captured and safely stored without causing any leakage or contamination issues.

References

Abid K, Gholami R, Elochukwu H, Masood M, Bing CH, Muktadir G (2018) A methodology to improve nanosilica based cement used in CO_2 sequestration sites. Petroleum 4:198–208. https://doi.org/10.1016/j.petlm.2017.10.005

Afroz R, Hassan MN, Ibrahim NA (2003) Review of air pollution and health impacts in Malaysia. Environ Res 92(2):71–77

Anwar MN, Fayyaz A, Sohail NF, Khokhar MF, Baqar M, Khan WD, Rasool K, Rehan M, Nizami AS (2018) CO_2 capture and storage: a way forward for sustainable environment. J Environ Manag 226:131–144. https://doi.org/10.1016/j.jenvman.2018.08.009

Araújo OdQF, de Medeiros JL (2017) Carbon capture and storage technologies: present scenario and drivers of innovation. Curr Opin Chem Eng 17:22–34. https://doi.org/10.1016/j.coche.2017.05.004

Arocho I, Rasdorf W, Hummer J (2014) Methodology to forecast the emissions from construction equipment for a transportation construction project, construction research congress 2014: construction in a global. Network:554–563

Bachu S, Gunter W, Perkins E (1994) Aquifer disposal of CO_2: hydrodynamic and mineral trapping. Energy Convers Manag 35(4):269–279. https://doi.org/10.1016/0196-8904(94)90060-4

Bachu S, Bonijoly D, Bradshaw J, Burruss R, Holloway S, Christensen NP, Mathiassen OM (2007) CO_2 storage capacity estimation: methodology and gaps. Int J Greenhouse Gas Control 1(4):430–443. https://doi.org/10.1016/S1750-5836(07)00086-2

Bennaceur K, Gielen D, Kerr T, Tam C (2008) CO_2 capture and storage: a key carbon abatement option. OECD

Chadwick R, Noy D, Holloway S (2009) Flow processes and pressure evolution in aquifers during the injection of supercritical CO_2 as a greenhouse gas mitigation measure. Pet Geosci 15(1):59–73. https://doi.org/10.1144/1354-079309-793

Choi M, Na K, Kim J, Sakamoto Y, Terasaki O, Ryoo R (2009) Stable single-unit-cell nanosheets of zeolite MFI as active and long-lived catalysts. Nature 461(7261):246–249

Cuéllar-Franca RM, Azapagic A (2015) Carbon capture, storage and utilisation technologies: a critical analysis and comparison of their life cycle environmental impacts. J CO2 Util 9:82–102. https://doi.org/10.1016/j.jcou.2014.12.001

Dindi A, Quang DV, Vega LF, Nashef E, Abu-Zahra MRM (2019) Applications of fly ash for CO_2 capture, utilization, and storage. J CO2 Util 29:82–102. https://doi.org/10.1016/j.jcou.2018.11.011

Han W, Kim K-Y, Esser R, Park E, McPherson B (2011) Sensitivity study of simulation parameters controlling CO_2 trapping mechanisms in saline formations. Transp Porous Media 90(3):807–829. https://doi.org/10.1007/s11242-011-9817-7

Herzog H (2009) Carbon dioxide capture and storage. https://sequestration.mit.edu/pdf/2009_CO2_Capture_and_Storage_Ch13_book.pdf

Herzog H, Drake E, Adams E (1997) CO_2 capture, reuse, and storage technologies for mitigating global climate change. A White Paper, 1–70

Iglauer S, Pentland CH, Busch A (2015) CO_2 wettability of seal and reservoir rocks and the implications for carbon geo-sequestration. Water Resour Res 51(1):729–774. https://doi.org/10.1002/2014WR015553

IPCC (2005) IPCC special report on carbon dioxide capture and storage. Prepared by Working Group III of the Intergovernmental Panel on Climate Change, Cambridge, UK and New York, NY, USA

IPCC (2007) Climate Change 2007: mitigation of Climate Change: contribution of working III to the fourth assessment report of the Intergovernmental Panel on Climate Change (Metz B, et al, eds). Cambridge/New York: Cambridge University Press

Irfan MF, Bisson TM, Bobicki E, Arguelles-Vivas F, Xu Z, Liu Q, Babadagli T (2018) CO_2 storage in saline aquifers by dissolution and residual trapping under supercritical conditions: an experimental investigation. Colloids Surf A Physicochem Eng Asp 548:37–45. https://doi.org/10.1016/j.colsurfa.2018.03.062

Jensen MD, Schlasner SM, Sorensen JA, Hamling JA (2014) Operational flexibility of CO_2 transport and storage. Energy Procedia 63:2715–2722. https://doi.org/10.1016/j.egypro.2014.11.294

Leung DY, Caramanna G, Maroto-Valer MM (2014) An overview of current status of carbon dioxide capture and storage technologies. Renew Sust Energ Rev 39:426–443. https://doi.org/10.1016/j.rser.2014.07.093

Lin B, Ahmad I (2017) Analysis of energy related carbon dioxide emission and reduction potential in Pakistan. J Clean Prod 143.(Supplement C:278–287. https://doi.org/10.1016/j.jclepro.2016.12.113

Moazzem S, Rasul MG, Khan MMK (2012) A review on technologies for reducing CO_2 emission from coal fired power plants, thermal power plants (Rasul M, ed). InTech. SPE-23641-PA.10.5772/31876

Orr FM Jr (2009) CO_2 capture and storage: are we ready? Energy Environ Sci 2(5):449–458. https://doi.org/10.1039/B822107N

Raza A (2017) Reservoir characterization for CO_2 injectivity and flooding in petroleum reservoirs, offshore Malaysia, Curtin Sarawak, Malaysia, Australia, PhD thesis, 231 pp

Raza A, Gholami R, Sarmadivaleh M, Tarom N, Rezaee R, Bing CH, Nagarajan R, Hamid MA, Elochukwu H (2016) Integrity analysis of CO_2 storage sites concerning geochemical-geomechanical interactions in saline aquifers. J Nat Gas Sci Eng 36PA:224–240. https://doi.org/10.1016/j.jngse.2016.10.016

Raza A, Meiyu G, Gholami R, Rezaee R, Rasouli V, Sarmadivaleh M, Bhatti AA (2018) Shale gas: a solution for energy crisis and lower CO_2 emission in Pakistan. Energy Sources, Part A 40 (13):1647–1656. https://doi.org/10.1080/15567036.2018.1486486

Raza A, Gholami R, Meiyu G, Rasouli V, Bhatti AA, Rezaee R (2019) A review on the natural gas potential of Pakistan for the transition to a low-carbon future. Energy Sources, Part A 41 (9):1149–1159. https://doi.org/10.1080/15567036.2018.1544993

Rutqvist J, Birkholzer J, Cappa F, Tsang CF (2007) Estimating maximum sustainable injection pressure during geological sequestration of CO_2 using coupled fluid flow and geomechanical fault-slip analysis. Energy Convers Manag 48(6):1798–1807. https://doi.org/10.1016/j.enconman.2007.01.021

Saxena N, Bhargava R (2017) A review on air pollution, polluting agents and its possible effects in 21st century. Adv Biores Soc Educ India 8(2):42–50. https://doi.org/10.15515/abr.0976-4585.8.2.4250

Songolzadeh M, Soleimani M, Takht Ravanchi M, Songolzadeh R (2014) Carbon dioxide separation from flue gases: a technological review emphasizing reduction in greenhouse gas emissions. Sci World J 2014:1–34

Varre SBK, Siriwardane HJ, Gondle RK, Bromhal GS, Chandrasekar V, Sams N (2015) Influence of geochemical processes on the geomechanical response of the overburden due to CO_2 storage in saline aquifers. Int J Greenhouse Gas Control 42:138–156. https://doi.org/10.1016/j.ijggc.2015.07.029

Wennersten R, Sun Q, Li H (2015) The future potential for carbon capture and storage in climate change mitigation–an overview from perspectives of technology, economy and risk. J Clean Prod 103:724–736. https://doi.org/10.1016/j.jclepro.2014.09.023

White MD, McGrail BP (2003) Numerical investigations of multifluid hydrodynamics during injection of supercritical CO_2 into porous media. In: Greenhouse gas control technologies-6th international conference. Elsevier, pp 449–455. https://doi.org/10.1016/B978-008044276-1/50072-6

Zhao X, Liao X, He L (2016) The evaluation methods for CO_2 storage in coal beds, in China. J Energy Inst 89(3):389–399. https://doi.org/10.1016/j.joei.2015.03.001

Chapter 2
Sources of Carbon Dioxide and Environmental Issues

M. N. Anwar, M. Iftikhar, B. Khush Bakhat, N. F. Sohail, Mujtaba Baqar, A. Yasir, and A. S. Nizami

Abstract The rapid increment in the anthropogenic activities has enhanced carbon dioxide (CO_2) emissions and has given birth to pressing environmental issues worldwide. CO_2 imparts a significant role in global warming that leads to global climate change. The increased dependency on fossil fuels, in the form of coal, oil and natural gas, has raised the concentration of CO_2 in the atmosphere from 280 ppm to 413 ppm. In the past decade, the CO_2 emissions were taking place at the rate of 2 ppm/year and has led several risks to human life including glacier melting, floods, heat waves, droughts, cyclones, hurricanes, and food security issues. Countries like China, United States, India, Russia, Japan, Korea, Germany, Iran, Canada, United Kingdom, and others contribute the lion's share in global CO_2 emissions.

Burning of fossil fuels adds around 6.5 billion tons of CO_2 in the atmosphere every year. In addition, ever growing population has exacerbated the deforestation activities, hence enhancing the CO_2 emissions. The population increased from around 1.65 billion in 1900 to nearly 7.4 billion in 2015. Overpopulation accelerate natural resources exploitation resulting in the utilization of fossil fuels at an alarming rate. Natural processes like forest fires and volcanic eruptions are also contributing to global CO_2 emissions. Consequently, the climatic shift induced extreme weather events have posed massive damages to planet earth and gravely affected the human life and biodiversity. Since 1960 the extent of weather-related natural disasters increased three times. These disasters have caused more than 60,000 deaths worldwide mainly affecting the developing countries. This chapter aims to pen down the major sources of CO_2 emissions and their environmental issues.

M. N. Anwar (✉) · M. Iftikhar · B. Khush Bakhat · M. Baqar · A. Yasir
Sustainable Development Study Center (SDSC), Government College University, Lahore, Pakistan

N. F. Sohail
Institute of Environmental Sciences and Engineering, School of Civil and Environmental Engineering, National University of Sciences and Technology, Islamabad, Pakistan

A. S. Nizami
Center of Excellence in Environmental Studies (CEES), King Abdulaziz University, Jeddah, Saudi Arabia

Inamuddin et al. (eds.), *Sustainable Agriculture Reviews 37*, Sustainable Agriculture Reviews 37, https://doi.org/10.1007/978-3-030-29298-0_2

Keywords Climate change · Greenhouse gases · Carbon dioxide emissions · Natural and anthropogenic activities · Extreme weather events · Global warming · Fossil fuels

Book Chapter Highlights

- The greenhouse gases are increasing with the passing years because of multiple factors.
- CO_2 is a primary greenhouse gas responsible for increase in the global temperature.
- CO_2's concentration has mounted to 400 ppm from 280 ppm value of pre-industrial era.
- Anthropogenic activities such as the burning of fossil fuels for energy production, industrial operations, and agricultural activities driven by overpopulation has caused increase in CO_2 concentration.
- Increased global temperature has led to multiple environmental issues like cyclones, tornadoes, hurricanes, floods, droughts, and agricultural impacts.

2.1 Introduction

Carbon dioxide (CO_2) is colorless, trace atmospheric gas, which despite its low concentrations plays a decisive role in the global climate and Earth life cycle (Ghommem et al. 2012; Shakerian et al. 2015). The atmosphere and its global warming phenomena such as capturing the ultraviolet radiations by the greenhouse gases, i.e., CO_2, water vapor, methane, nitrous oxide, and chlorofluorocarbons are what makes planet earth livable by helping it in maintaining enough warm climate. This balance has helped humanity to survive, but more recently, a significant change has been recorded in the concentrations of greenhouse gases in general and CO_2 in particular.

CO_2 emissions from multiple sources, mainly industrial operations, and energy production are playing a vital role in determining the global climate. The concentration of CO_2 has now mounted to 400 ppm from 280 ppm leading to a 0.8 °C rise in global surface temperature (Pachauri et al. 2014). Climate Change has relatively exacerbated more in past 30 years and the Intergovernmental Panel on Climate Change's (IPCC), with high confidence, has declared the period between 1983 and 2012 as the warmest in past eight centuries (Ho 2018).

This increased temperature has increased the frequency and scale of extreme weather events such as droughts, floods, wildfires, heat waves, and hurricanes. Sea level has risen at a rate of 0.8 mm/year from 1971 to 2010. This has been accompanied by the glaciers' retreat, at 226 Gt/year the 40 years above and at 301 Gt/year between 2005 and 2009. The number of warm days and warm nights have also mounted further in the last half of the previous century. Resultantly, more and intense heat waves are occurring across the globe in general and at Europe, Asia and Australia in particular (Ho 2018; Kamaruddin et al. 2016).

The hot weather has also intensified the frequency of category four and five hurricanes (Bender et al. 2010; Dinan 2017; Lee and Ellingwood 2017). In 2012 alone, hurricanes caused 52 billion USD loss to the USA with an economic toll as high to 5% of the global GDP, as suggested by some studies (Dinan 2017). The extreme events' occurrence has increased to four to fivefolds globally since the 1980s. Droughts, exacerbated and intensified by climate change, too have caused significant damage to the world's economy and have afflicted grave consequences over poorer nations such as 2011, and 1990–1991 drought episodes at Somalia and Zimbabwe. These extreme events are giving birth to other serious issues such as climate refugees, and food insecurity. Already one million Bangladesh population has been displaced owing to floods, and countries like Pakistan are facing food insecurity due to climate change aftermaths. Pakistan faced severe droughts from 1999 to 2003 and floods from 2010. These events have led Pakistan to endure massive economic and socio-cultural impacts. Earth's Warming by 3 °C can reduce the food yield at Asia – which supplies 2/3rd of world's food demand – by one fourth, severely affecting the food availability globally (Anwar et al. 2018).

There have been concerns and widespread worries about the changing climate, both on regional and global scales, especially in the late twentieth century to date. There has been a unanimous opinion of scientists claiming the rapid increase in the greenhouse gases the primary reason behind this undesired climate change. CO_2 is a greenhouse gas which is believed to have the lion's share, nearly three quarters, in global greenhouse gas emissions and it is playing a significant role in altering the global temperature (Azevedo et al. 2018). Hence CO_2 is playing a dominant role in the greenhouse effect. The primary reason behind the elevation of this highly impactful gas is increased consumption of fossil fuels (Garnier et al. 2019).

Hence, all these adverse effects and future projections call for curbing these emissions. This can be a daunting task, but collectively the world can get the job done. IPCC suggests that the greenhouse gases emissions must be curbed by 50–80% by 2050 in order to prevent climate inflicted a disastrous collapse of planet Earth. United Nations, in the bid to save planet, gathered almost 190 heads of states at Paris for a 21st conference of parties in 2015 (COP 21) and consensus was reached to limit the rise of average global temperature below 2 °C by the end of this century employing multiple approaches: using clean and renewable energy, afforestation, energy conservation and carbon capture and storage (CCS) techniques (Anwar et al. 2018).

Although there are many sources of CO_2 emissions; however, the most significant contributions come from anthropogenic activities such as combustion of various forms of fossil fuels such as coal, oil and gas. The graph below represents the volumes of CO_2 emissions from different sources. The major contributor is coal followed by oil and gas (Ritchie and Roser 2017) (Fig. 2.1).

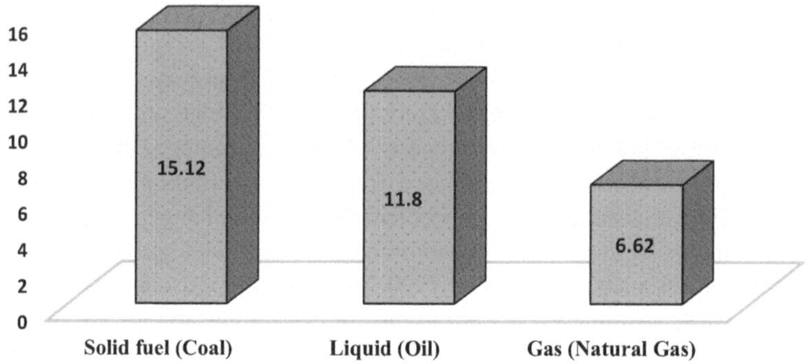

Fig. 2.1 World carbon dioxide emissions from coal, oil and natural gas measured in billion tonnes in the year of 2013 (Boden et al. 2017; Ritchie and Roser 2017)

2.2 Source of Carbon Dioxide

2.2.1 Anthropogenic Activities

Scientific literature denotes that the increase in atmospheric CO_2's concentration is because of the overwhelming human activities (Letcher 2018). The CO_2 concentration has mounted from 280 ppm (before the Industrial Revolution) to 413 ppm (observed on April 26, 2017) increasing the average global temperature (Letcher 2018). There are multiple anthropogenic activities responsible for this elevation; however, the combustion of fossil fuels for various purposes could be a primary culprit behind this increase from 1920 until present (Le Quéré et al. 2012).

Figure 2.2 shows the increase in the global monthly mean CO_2 concentrations over the years 1960–2017. It is evident that the concentration of CO_2 has registered a sharp increase throughout decades as depicted by the significant difference in the concentrations of 1983 and pre-industrial era, 340 ppm, and 270 ppm respectively. This shows around 26% increase in the concentration of CO_2 from 1850 to 1983 (LaMarche et al. 1984). The situation is more alarming if we compare the concentration of CO_2 from the preindustrial era (270 ppm) to 2020 (around 414 ppm) where an approximate 53% rise can be observed. This also indicates the share of industrial activities in the rising level of CO_2 concentration (Ritchie and Roser 2017).

2.2.1.1 Carbon Dioxide Emissions from Fossil Fuels' Combustion

In the pre-industrial era, the concentration of CO_2 in the atmosphere was very low. The CO_2 concentration has remained constant and fluctuated within a very small range for previous 800,000 years and remain within the range of 170–300 ppm; however, a gradual fluctuation has been recorded throughout the postindustrial era.

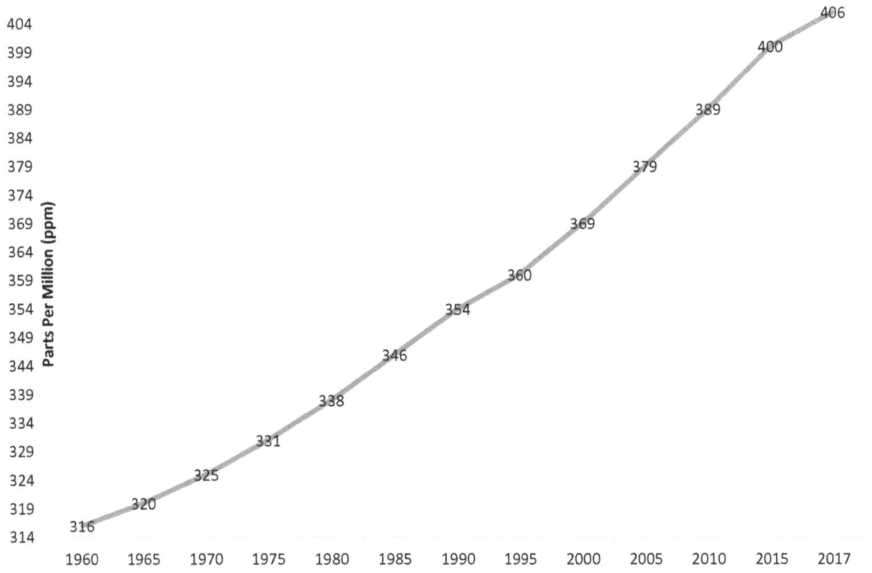

Fig. 2.2 The global CO_2 atmospheric concentration. The global mean annual concentration of CO_2 emissions are measured in parts per million (ppm) (Ritchie and Roser 2017)

In just a few years it is expected that the emissions will rise to 100% to that of pre-industrial era's level (Luthi et al. 2008; Tans 2008; IIASA).

In the last decade, CO_2 concentration increased at the rate of 2 ppm/year. The Intergovernmental Panel on Climate Change's (IPCC) fifth assessment report indicates that the increase in CO_2 concentration can be attributed to several anthropogenic activates. These activities are mainly driven by ever-growing energy demand (Victor et al. 2014).

Every year humans dispose of tons of CO_2 in the atmosphere. A significant share of this emission, about 6.5 billion tons, comes from the burning of fossil fuels and deforestation produces the remaining 1.5 billion tons. Among the top CO_2 emitters are China, United States, India, Russia, Japan, and Canada. The emissions activities like combustion of fossil fuels, cement manufacturing, and gas flaring are primary sources of CO_2 at these countries. For instance, the power plants at the USA act as a primary driver behind climate change owing to the lion's share, almost one third, in CO_2 emissions. Same is the case with India which is becoming the wellknown country for its CO_2 emissions. The emissions by India are reported to be 2000 million tons, and the power sector of this country is contributing nearly half of the country's carbon emissions (Chandel et al. 2016).

It is observed that 90% of emissions from the energy sector is from CO_2 alone. The rest 10% comprises methane (9%) and nitrous oxide (1%). The chart shows the sectors contributing to global greenhouse gases' emissions in which the energy sector dominates the share with approximately 67.33% (Fig. 2.3). Economic growth and developmental activities increase the demand for energy. This global energy

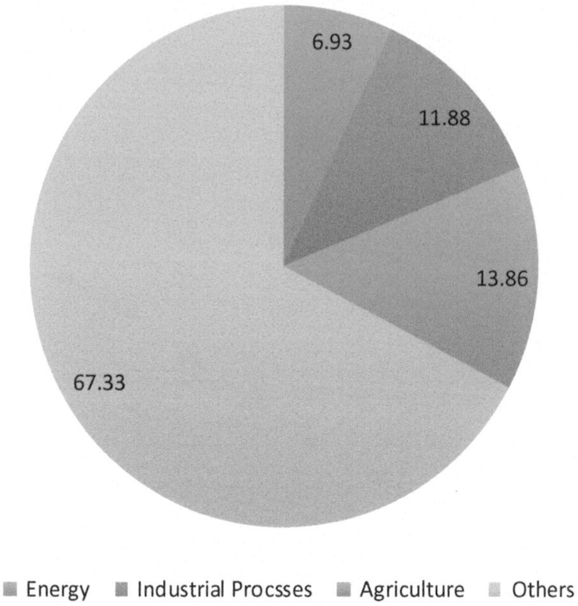

Fig. 2.3 Shows the estimated percentage share of total global anthropogenic activities in greenhouse gases in 2014. Others include large-scale biomass burning, post-burn decay, peat decay and indirect emissions (IEA Report 2017)

demand increased by 150% from 1971 to 2015. The world's primary energy supply is fulfilled mainly by fossil fuels, and these sources accounted for 82% of energy supply in 2015. The increase in fossil fuel combustion can be linked to the rise in the world's energy demand causing annual CO_2 emissions to reach over 33 Gt CO_2 (2015) (IEA Report, 2017).

2.2.1.2 Trends in emissions

As mentioned above too, the emissions from fossil fuels' combustion have the lion's share in the greenhouse effect and climate change (Wang et al. 2018). The emissions of ten countries such as China, United States (US), India, Russia, Japan, South Korea, Germany, Iran, Canada, United Kingdom (UK) and others alone constitute around two-thirds of global greenhouse gases' emissions (Fig. 2.4) (Nejat et al. 2015). The records demonstrate that the US and China are major contributors of total CO_2 emissions (Azevedo et al. 2018). In 2016, these both countries accounted for 43.3% of global CO_2 emissions (Wang et al. 2018).

The worldwide economic growth and development causes increased demand for energy. Total primary energy supply (TPES) is a measure of global energy demand. According to an International Energy Agency report, from 1971 to 2015 there is a significant rise in total primary energy supply and the increase of 150% is observed between these years. This energy mostly depends upon fossil fuels. (IEA Report 2017).

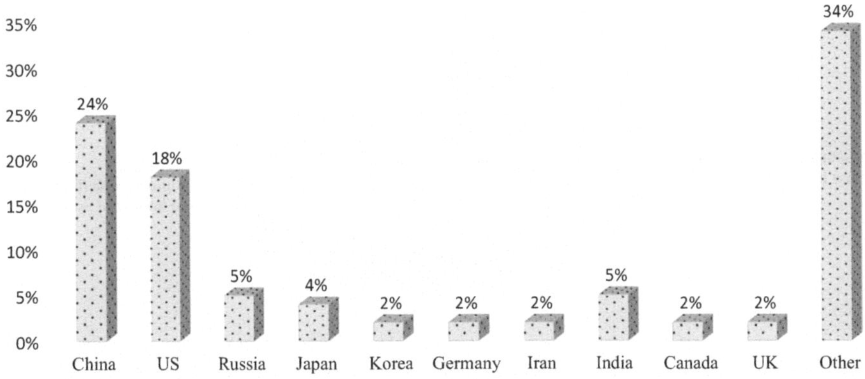

Fig. 2.4 Indicates the percentage share of the top ten countries in global CO_2 emissions in 2010 (Nejat et al. 2015)

Table 2.1 Shows the percentage share by fuel in 2015. The total world primary energy supply and carbon dioxide emissions

Carbon dioxide emissions (CO_2)	Shares by fuel	Total primary energy supply (TPES)
45%	Coal	28%
34%	Oil	32%
20%	Gas	21%
1%	Others	19%

Others include the nuclear, hydro, solar, biofuels, waste, tide, wind and geothermal. (IEA Report 2017)

In 2015, coal represented 28% of the world total primary energy supply but it contributed 45% of the total global carbon dioxide emissions. The reason behind the large percentage of carbon dioxide emissions is the carbon content per unit of energy released. The Table 2.1 above shows that after coal, oil accounted for 32% of the total primary energy supply and its percentage emission of carbon dioxide is 34%. Gas contributed to 21% of total primary energy supply and its carbon dioxide emissions were 20%, far less behind the emissions from coal. (IEA Report, 2017).

The IPCC Report proposed the change in the greenhouse gases' emissions between 1970 and 2010 and the relative drivers for these changes. It can be inferred from the report that the major concern is CO_2's emissions, produced directly from fossil fuels burning and industrial processes. An increase of 10% in the CO_2 emissions due to fossil fuel burning and industrial processes can be observed in the last 40 years since the value rose to 65% from 55%. Although the emissions from CO_2 FOLU (forestry and other land use) reduced from 17% (1970) to 11% (2010) but the total CO_2 emissions escalated by 4%, as it is mounted to 76% from 72% (Victor et al. 2014). Therefore, the report indicates the stack of CO_2 emissions in the overall greenhouse effect. In this way, to achieve sustainable development, we needs to take care of these CO_2 emissions. It can be possible if we employ some techniques like carbon capturing and storage (Anwar et al. 2018, Victor et al. 2014) (Fig. 2.5).

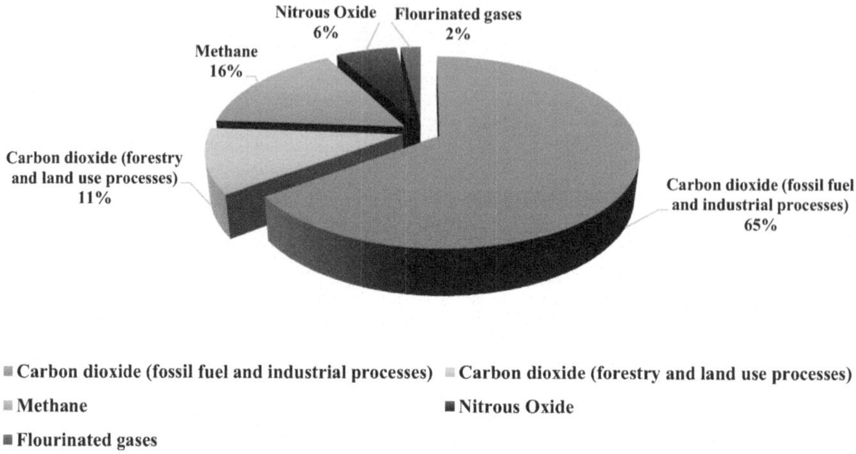

Fig. 2.5 The figure shows the total contribution of greenhouse gases emissions in 2010. The carbon dioxide emissions are more than collective emissions of methane, nitrous oxide and fluorinated gases (Victor et al. 2014)

In 2010, fossil fuel emissions caused 65% release of CO_2 in the atmosphere. Forestry and land use are also the drivers behind CO_2 emissions and contributed 11% of total Global GHG emissions. The concentration of methane, nitrous oxide, and fluorinated gases some other greenhouse gases, remained very low as compared to CO_2 alone (Victor et al. 2014). The Table 2.2 shows the global CO_2 emissions from fuel combustion between 1971 to 2015.

If we calculate the total CO_2 emissions of every country of the world starting from 1750 to 2016, then we can measure the cumulative emissions of every country till date. It is observed that the emissions were not high before 1750. UK (United Kingdom) first started anthropogenic emissions by operating its industries through fossil fuels combustion followed by North America. Later Asia, Africa, and Latin America also joined these countries in the bid to increase global warming. The recent data shows that U.S and Europe still dominate in greenhouse gases' emissions; however, China's boost in the industrial sector has also made its second-largest cumulative emitter (Ritchie and Roser 2017).

2.2.1.3 Industrial Emissions

The industrial sector has its large share in greenhouse gas emissions mainly CO_2 (Mardones and Flores 2018). The nations argue that energy consumption is mandatory for strong economic growth. In this way, the combustion of fossil fuels for energy production is believed to be very vital (Wang et al. 2018). The Intergovernmental Panel on Climate Change (IPCC) report of 2015 shows that 21% of greenhouse gas emissions are directly coming from the industrial sector. (Mardones and Flores 2018).

Table 2.2 Shows the global carbon dioxide emissions in million tons from fuel combustion, 1971–2015

Million tons of carbon dioxide emissions (World)											
Source	1971	1975	1980	1985	1990	1995	2000	2005	2010	2014	2015	Percentage Change
Fuel Combustion	13942.2	15484.1	17706.3	18246.5	20,509	21,365	23144.3	27,045	30434.4	32324.7	32294.2	57.50%
Coal	5229.3	5648	6600.8	7394.2	8286.2	8505.9	8962.6	11453.7	13725.6	14834.2	14512.7	75.10%
Oil	6668.1	7581.5	8389.8	7772.9	8499.2	8792.1	9531.7	10,286	10535.2	10,958	11169.1	31.40%
Natural gas	2043.6	2249.4	2709.2	3070.1	3679.9	3984.6	4550.2	5206.4	6026	6358.3	6437	74.90%
Marine Bunkers	353.79	341.07	357.33	306.76	371.64	428.54	498.37	571.92	662.94	627.73	657.04	76.80%
Aviation Bunkers	169.22	173.87	202.13	224.9	258.94	290.34	355.82	422.78	457.66	504.49	529.69	104.60%

The fuel combustion comprised of carbon dioxide emissions from different sources such as coal, oil, natural gas, marine bunkers and aviation bunkers. (IEA Report 2017)

Fig. 2.6 Shows the share of total final energy consumption (TFEC) in China, 2011. The share of energy consumption by number of industries, the overwhelming transportation use, high demand of energy by residential sector, commercial sector and others. (Nejat et al. 2015)

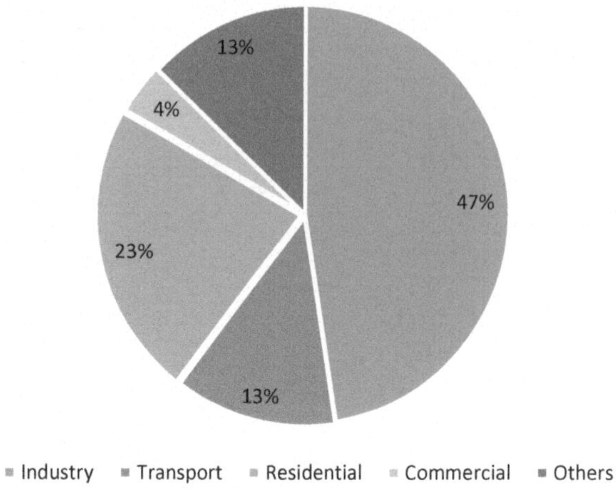

The major types of industrial operations responsible for a large chunk of greenhouse gases emission could be divided into three categories: chemical handling and processes, energy generation and waste management (Mardones and Flores 2018).

With the increased industrialization, urbanization flourished, leading to rapid development of infrastructures such as buildings and roads. Cement industry flourished to cater to the demand for cement mounting day by day without considering its environmental impacts (Chandel et al. 2016).

China is leading the developing nations since 2005 as CO_2 emitter. The main reason behind the hike in emissions has increased energy consumption by different sectors of the country (Pappas et al. 2018). In 2011, China's emissions of CO_2 accounted for nearly 24% of total world emissions and in 2010 heavy industry and economic growth become a reason behind China's elevated emissions. In 2007 China took over the U.S as the world's largest emitter of carbon dioxide. (Nejat et al. 2015) (Fig. 2.6).

2.2.1.4 Overpopulation and Carbon Dioxide Emissions

Overpopulation has been directly linked with the increase in global CO_2 emissions. A direct relation can be found between the increase in the population of our planet and CO_2 emissions. Population growth plays a major role in natural resources' exploitation, hence laying stress on the environment (Anqing Shi 2003). It is an increasing concern that population growth is one of the major driving forces behind the rapidly increasing global carbon dioxide emissions. (Shi 2003).

The population change causes great impact on environmental stress (Shi 2003). A large population demands more energy consumption and more resources are exploited to fulfill the needs of population (Wang et al. 2018). A Table 2.3 display that the carbon dioxide emissions grew by 26.88% in developed countries for the years 1975 and 1996 period, while population increased 16.38% during the same

Table 2.3 Growth of carbon dioxide emissions and population sample of 93 countries during the time period of 1975–1996

Countries	Developed countries	Developing countries	Developed countries	Developing countries
Factors	CO_2 emissions (million metric tons of carbon)		Population (million)	
1975	2191	952	696	2763
1996	2780	2286	810	4130
Percentage increase	26.88	140.13	16.38	49.82

The table shows that the emissions of carbon dioxide in million metric tons of carbon by population of developed and developing countries. The percentage increase of carbon dioxide emissions and population by developed and developing countries has also been presented. (Shi 2003)

period. In developing countries, emissions grew by 140.13%, whereas population increased by 49.48%. (Shi 2003) (Table 2.3).

2.2.1.5 Agriculture Sector

The use of fossil fuels in the agriculture sector is related to the use of machinery in farmland, the manufacturing of fertilizers, and trade of feeds. This consumption of fuel emits direct CO_2 gas in the atmosphere and ultimately becomes the source of emissions. The fossil fuel driven practices and mechanization in the agriculture sector cause a hike in greenhouse gases emissions (Fig. 2.7) (Garnier et al. 2019). According to the U. S emission inventory around 8% of the greenhouse gas emissions are from the agricultural practices such as consumption of fossil fuel for water pumping, nitrogen-rich fertilizer, and irrigation (Waheed et al. 2017).

The agricultural emissions have been increasing with time. A growing population and deforestation are the two main reasons behind the enhancement of increasing emissions in the agriculture sector. The increase in demand for food converts the forest land into agricultural land and results in the release of stored CO_2 in trees and soils (Ritchie and Roser 2017).

2.2.2 *Natural Sources of Carbon Dioxide*

2.2.2.1 Forest Fires

Forest fire causes detrimental effects on the environment. Its negatives impacts are long lasting (Rahman et al. 2018). Forest is believed to be a good sink of CO_2, but it also acts as a source of CO_2 emissions when it catches fire. Initially, the forest fires were natural, and related impacts were less because of low frequency. It is expected that the frequency and severity of wildfires will increase in the future due to human-induced climate changes such as increased temperature, dry conditions, humidity

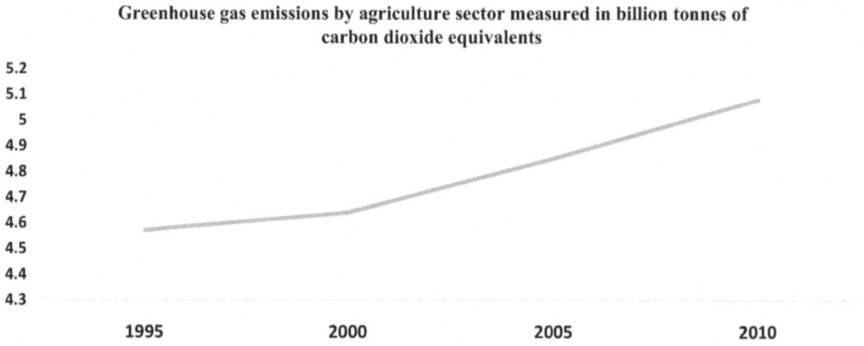

Fig. 2.7 Greenhouse gas emissions by agriculture sector measured in billion tonnes of CO_2 equivalents. CO_2 equivalents measure the total greenhouse gas potential of the full combination of gases, weighted by their relative warming impacts (Ritchie and Roser 2017)

Table 2.4 The estimation of regional wildfire carbon emissions from forest fires in Alaska

Region	Year	Teragram (Tg) Carbon per year
Yukon River Basin	2004	81
Alaska boreal forests	2004	69
Alaska boreal forests	2000–2009	14
Alaska boreal forests and wetlands	1950–2009	39

In Alaska, from over the past 50 years the rate and severity of wildfires of boreal forest increased enormously resulting in rise of carbon emissions into the atmosphere. (Potter 2018)

level, and thermal inversions (Paula García-Llamas et al. 2019). Hence, the anthropogenic activities are exacerbating the rate of natural forest fires (Waheed et al. 2017). The Table 2.4 shows the rate and severity of wildfires of boreal forest.

Another source of CO_2 emission in the forest is the exposed soil. When a large number of trees and plants are removed, then soil microbes release the stored CO_2 of dead matter into the atmosphere, so it becomes the direct source of CO_2 emissions too (Waheed et al. 2017).

2.2.2.2 Volcanic eruption

Volcanic eruptions are among the primary sources of carbon emissions, and their number is increasing from the past few years. It is studied that from the Holocene period, each year almost 50–70 volcanoes erupted from the Earth. There are also carbon emissions from nearly 500 volcanoes; these degassing emissions are taking place from either the hydrothermal system or open-vent system. The Table 2.5 shows the CO_2 soil gas flux (t/year) from historically active volcanoes in various countries (Burton et al. 2013).

The dissolved CO_2 in the layer of the mantle, the CO_2 from sub-ducted crustal substance and the de-carbonization of less dense crustal substances are among the

Table 2.5 Diffused CO_2 from historically active volcanoes

Volcano	Country	CO_2 soil gas flux (t/year)
Erebus	Antarctica	14,600
Nea Kameni	Greece	5621
Hengill Volcanic System	Iceland	165,345
Krafla Geothermal System	Iceland	84,000
Merapi	Indonesia	78,475
Etna	Italy	1,000,000
Usu	Japan	60,712
Cerro Negro	Nicaragua	1,022,000
Lassen	USA	35,000
Ukinrek Maars	USA	11,863

The total carbon dioxide flux measured at each volcano in various countries (Burton et al. 2013)

direct emissions of active volcanoes. Geologically the Volcanic and tectonic structures play a significant role in global CO_2 emissions. The studies show that the subaerial volcanism emits around 300 Mt/year of CO_2. There is also a gaseous non-volcanic inorganic source adding around 300 Mt/year of CO_2 emissions into the atmosphere by the tectonically active areas (Chiodini 2005).

2.3 Environmental Issues Related to Carbon Dioxide Emissions

As discussed in the start of this chapter that exacerbated carbon dioxide emissions lead to an unprecedented rise in overall earth's temperature leading to more frequent and widespread incidents of climate-induced extreme events such as heat waves, droughts, floods, hurricanes, tornadoes, and irregular rainfall patterns. It has been proved that these conditions are due to anthropogenic activities (Stott et al. 2004; Pall et al. 2011).

Many anthropogenic activities cause the phenomenon of climate change to occur. Climate change ultimately become a reason of high extent of global extreme events. From past years humans faced the extreme conditions in many sectors and the major sectors are agriculture, infrastructure and health. This makes the very diverse range of extreme events which differ due to the climatic conditions (Peters et al. 2001).

Some of the extreme weather events due to climate change are discussed below:

2.3.1 Cyclones and Hurricanes

Most of the cyclones occur in most tropical regions like oceans and are called tropical cyclones. These tropical cyclones pose a serious threat to the population and infrastructure on the coastal areas and the marine activities like shipping and other offshore businesses. It is estimated that almost 90 tropical cyclones occur globally every year and this number of cyclones has almost remained constant over the years. Although the global occurrence of cyclones in the tropical oceans is almost constant, but there can be the variability of cyclones occurrence at inter-annual to the multi-decadal time interval in basins of the individual ocean (Webster et al. 2005). This variability in the occurrence of the cyclones at regional levels in combination with inter-annual and multi-decal variability gives a picture of future disaster occurred due to the cyclones in specific regions (Kossin et al. 2010).

Several studies have estimated that, along with the natural variability in the sea surface temperature there is noticeable increase in the warming trends of the tropical oceans owing to the increasing global warming in general and carbon dioxide in particular (Karoly and Wu 2005; Knutson et al. 2010; Gillett et al. 2008). It is estimated that the anthropogenic activities caused the warming of the North West Pacific and North Atlantic regions through the twentieth century (Handmer et al. 2012).

Statistical data below shows the number of cyclones, the number of countries hit by the cyclone and the number of disasters from the cyclones from 1970 to 2009. This can be seen easily from these graphs that although the number of cyclones has decreased from 2000 to 2009, but the number of disasters from these cyclones has exploded. This shows that there is a prominent increase in the intensity of these cyclones that have caused almost 40 times more disasters as compared to 1970 (Handmer et al. 2012) (Fig. 2.8).

In the following table, some of the major cyclones are listed along with the category and the year of occurrence (Table 2.6).

The increased level of the carbon dioxide has resulted in the triggered increase in the disasters due to the climate change such as hurricanes, cyclones and droughts, etc. Global warming has resulted in the increase in the temperature of the sea surface and thus the occurrence of the hurricanes and cyclones of the category four and category five especially in the tropical oceans (Anwar et al. 2018). The table below show list of some of the major hurricanes with their category and the year of the occurrence (Tables 2.7 and 2.8).

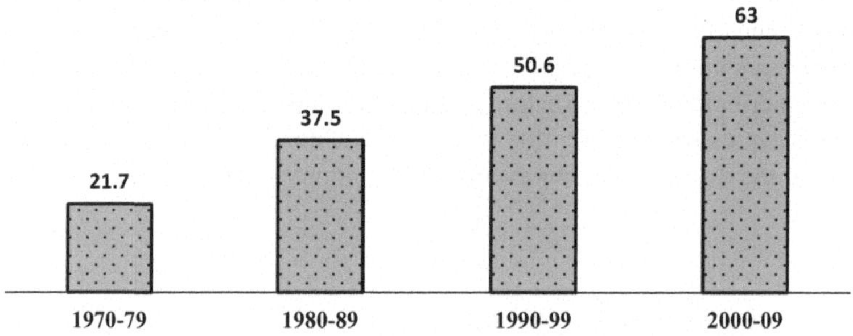

Fig. 2.8 Number of disasters triggered by tropical cyclones from 1970 to 2009. This data shows that the number of disasters triggered by the cyclones increased almost three folds in the year 2009 as compared to the number of disasters triggered by the cyclones in the 1970

Table 2.6 Some of the major cyclones of the world

South Atlantic Ocean			
Deni	2016	Subtropical storm	Reboita et al. (2019)
Eçaí	2016	Subtropical storm	da Rocha et al. (2018)
Guará	2017	Subtropical storm	da Rocha et al. (2018)
South West Indian Ocean			
Fantala	2015–16	Very intense tropical storm	Duvat et al. (2017)
North Indian Ocean			
Chapala	2015	Sever cyclonic storm	Sarker (2018)

The table shows the category of the storm and the year of its occurrence. Scale for the category of the cyclones/hurricanes is listed in the table number eight

Table 2.7 In this table, some important hurricanes are listed along with the causalities and category (Anwar et al. 2018)

Hurricane	Year	Category	Causalities	Location
Andrew	1992	5	44	Southern Florida
Katrina	2005	3	1833	Gulf of Mexico
Wilma	2005	5	63	Mexico, Cuba, Florida
Irene	2011	3	37–49	North Carolina
Sandy	2012	3	147	Cuba, The Dominican Republic Jamaica Haiti, USA, Ontario
Patricia	2015	5	2 Direct, 4 Indirect	South of Mexico
Hermine	2016	1	1	Florida
Otto	2016	3	18	Northern Costa Rica and Southern Nicaragua
Harvey	2017	4	77	Texas

Table 2.8 This table shows the Saffir-Simpson hurricane scale (Dolan and Davis 1992) used for the classification of the hurricanes in their respective categories

Category	Storm surge (m)	Maximum 1 min wind speed (ms^{-1})	Damage
1	1.2–1.5	33.0–42.4	Minimal
2	1.8–2.4	42.9–49.2	Moderate
3	2.7–3.7	49.6–58.1	Extensive
4	4.0–5.5	58.6–69.3	Extreme
5	> 5.5	> 69.3	Catastrophic

2.3.2 *Droughts*

These extreme events can be defined in different terms and perspectives depending on the stakeholders affected by it directly or indirectly. However, the scientific community defines drought as a prevailing condition of deficient precipitation, minimum soil moisture, a minimum flow of water in lakes and extremely low groundwater levels (Heim Jr 2002). Some of the important droughts are listed below in the table along with the year of their occurrence. Recent drought in Africa

Table 2.9 In this table, some of the major droughts of the world are listed along with the year of their occurrence

Year	Drought	Reference
1984	Ethiopia drought	Viste et al. (2013)
2005	Brazilian drought	Smith et al. (2014)
2010	Brazilian drought	Smith et al. (2014), (2014)
2011	Horn Of Africa drought	Viste et al. (2013)

was considered as the most severe drought in the past 60 years, resulting 380,000 people homeless and causes almost ten million people to be dependent on the human welfare aid (Anwar et al. 2018) (Table 2.9).

2.3.3 Heat Waves

Heat waves like other extreme events can be defined easily depending on the situation of the stakeholders involved directly or indirectly. Heat waves can be defined as the condition of excessive hot temperature with high humidity especially in the areas where water bodies are present. Heat waves are the extreme events which can affect the ecosystem directly and pose a serious threat to the species survival by decreasing the water availability at a higher level. It is predicted that in the future, with the increasing temperature, the number of heat waves would also increase. Not only the number of heat waves would increase, but intensity and duration of heat waves would also increase causing severe outcomes (Russo et al. 2014).

This warming, due to the heat waves, can decrease the net carbon dioxide exchange in the ecosystem generating the conditions for the droughts, and frequent heat waves decreasing the annual uptake of the carbon dioxide by the terrestrial ecosystems (Arnone Iii et al. 2008). It is predicted that owing to the increase of a number of heat waves carbon sinks in the ecosystem like trees will be converted into the carbon sources. During the heat wave in Europe in 2003 almost net source of 0.5 Pg (Picogram) Carbon per year of carbon dioxide was added to the ecosystem and reversed the process of 4 years of carbon sequestration (Ciais et al. 2005). Impacts of heat waves are not only determined by the warming of the ecosystem but are also measured by the organisms' sensitivity towards its ability to adopt the changing environmental harsh conditions (Huey et al. 2009). It is projected by the models and the observations that there would be an intensified increase in greenhouse gases over Europe and North America region which would cause severe heat waves in the future in this region (Meehl and Tebaldi 2004).

Impacts of heat waves are not only determined by the warming of the ecosystem but are also measured by the organisms' sensitivity towards its ability to adopt the changing environmental harsh conditions (Huey et al. 2009). It is projected by the models and the observations that there would be an intensified increase in greenhouse gases over Europe and North America region which would cause severe heat waves in the future in this region (Meehl and Tebaldi 2004). South Asia consists of

Table 2.10 In this table, some of the record-breaking heat waves events in various locations are listed along with the year of their occurrence. (Russo et al. 2014)

Year	Location
1980	United States
1994	Benelux
2003	Europe
2007	Greece
2010	Russia
2011	United States
2012	United States

one-fifth of the total world human population. This is a region which is more vulnerable to extreme events caused by climate change. Heat waves cause detrimental impacts on developing countries. In the regions of countries like Pakistan and India the combined effects of high temperature and humidity observed by deadly heat waves. The enormous number of humans and livestock death is caused by severe heat waves. In 2015, the deadliest heat wave was recorded in Pakistan and India around 3500 people were dead. (Im et al. 2017) (Table 2.10).

2.3.4 *Food System and Food Security*

Food security and production are also affected by extreme events. These impacts include a shortage of food or increase in the prices of the food that could be a great challenge especially for the poor developing countries (FAO 2008). This increase in prices badly influences the countries with low income where people spend most of their livelihood on food (OECD-FAO 2008). Extreme events related to weather directly affect the agriculture, its activities and products and the economies of the most of the developing countries of the world depends heavily on the agriculture (Easterling and Apps 2005; Easterling et al. 2007). Farmers are greatly affected by the weather and its climate events. For example, most of the farmers in Africa eat what they produce and sell only a small portion of it. In Kenya, all households grow maize, and almost 36% sell it (FAO 2009). In such conditions, the stakeholders (farmers and their governments) affected by the climatic conditions have very less or limited capacity for recovery (Easterling and Apps 2005).

Lobell et al. (2011) shared that yield production is decreased due to the warming conditions. It was estimated that the yield was reduced due to the warming of climate as compared to previous decades. Müller et al. (2011) evaluated in its project that there is a negative impact on the crops' production due to climate change. Falloon and Betts (2010) yielded that the increasing droughts and the floods affect the agricultural activities and there is a requirement of the mitigation practices to cop up with these extreme events.

If agricultural products are not consumed or used at the point of production, then these have to be transported to other places and processed and stored prior to transportation. This involves a complete systematic process from the production to the transportation involving the storage. All this chain of the food supply is exposed to multiple types of hazards. Almost at every step of this system, especially at the time of transportation, the food is exposed directly to extreme events. This makes the whole system of the food production, storage, and transportation to a risk of disturbance resulting from error or damage or blockage at any certain step of this system (IPCC 2014).

2.3.5 Glaciers Melting

Many of the high mountain ranges are situated at the margins of the tectonic plates increasing the potential of the hazards because of the climatic and geological situations. These mountains are exposed to many disasters like massive land sliding, ice cap melting, mountain permafrost degradation, flooding, avalanches and precipitation (Liggins et al. 2010; McGuire 2010). It is also said that the twentieth-century glacier melting is due to the 1970 post warming (Lemke et al. 2007). The most severe and dangerous change in the mountain environment is the glacier melting, and this change is also due to the anthropogenic activities (Paul et al. 2004). Alpine glaciers were in the maximum strength on this planet at the end of the ice age (Leclercq et al. 2011) but now these are depleted mostly due to the anthropogenic activities at a very rapid rate in the past some decades (Zemp et al. 2009). Many of the glaciers have retreated since the middle of the nineteenth century (Francou et al. 2000; Cullen et al. 2006). The rate of this degradation or melting has increased historically and has become more prominent in the twentieth century (Reichert et al. 2002; Haeberli and Hohmann 2008). It is also said that due to the melting of the glaciers owing to the increase in the earth's global temperature the lakes are expanding.

Most of the glaciers have reduced in size since 1850, and this shrinkage is because of global warming. This glacier melting has noticeable impacts both at the global and regional scales. For example, for the glacier melting there is a considerable increase in the sea level and this increase in sea level is a threat not only to the coastal and Island areas but to it also possesses dangerous disasters like floods, lake outbursts, glacial debris flow. These disasters affect social development, economic progress and ecological systems (Yang et al. 2015). For a short period, there are the advantages of the glacier melting like an increase in the river, lake flow that further makes the water available for different purposes. However, this glacier melting has very severe long-term effects, like this may decrease the water availability in a long-term perspective. This will also affect the food and crop production along with economic development especially in the regions at high altitude like Asia. Glaciers are essential natural resources for the living organism in the world. These not only provide water for different purposes but also act as the source of the buffer for the dry seasons and conditions such as droughts. Glaciers are very sensitive to global

Table 2.11 In this table, some of the glaciers of the Himalaya Mountains are mentioned along with the decrease in their area and the time duration of their decrease (Li et al. 2018)

Himalaya Mountains		
Name	Decrease	Time span
Kangwure Glacier	-2.70 M year $^{-1}$	2007–2010
Naimona'nyi Glacier	-1.11 M year $^{-1}$	2008–2013
Mapam Basin Glacier	-0.36 Km^2 year $^{-1}$	1999–2003

Table 2.12 In this table, some of the glaciers of Tibetan Plateau are mentioned along with the decrease in their area and the time duration of their decrease (Li et al. 2018)

Tibetan Plateau		
Name	Decrease	Time span
Zhadang Glacier	0.64 Km^2	1970–2007
Gurenhekou Glacier	0.23 Km^2	1970–2007
Baishui Glacier no.1	-2.70 M year^{-1}	1997–2004
Hailuogou Glacier	-6.16 M year^{-1}	1990–1997
74 Glaciers in Gongga mountains	29.2 Km^2	1966–2009

warming because a slight increase in the temperature makes the snow to melt due to these many glaciers in different regions are decreasing and retreating. This brings high risks to water resources that are the keystone resources for the sustainable development of environment, economy, and society. It is expected that the glaciers would lose their almost 54% area in 2050 and almost up to 80% part in 2100. Then according to this trend, the volume of ice would decrease by 79% in 2050 and by 92% in 2100. It a reported that the tipping point, which is the melting point of the ice of the glaciers would reach at its peak by 2020 and this conversion of ice in water would provide a large amount of water for the short period, but it would significantly decrease the runoff water in long term perspectives. These results are quite alarming and call for urgent attention to the mitigation measures to assure the long term water availability in the respective regions (Gao et al. 2018).

In the table numbers ten and eleven, some of the glaciers of the Himalaya mountain and the Tibetan Plateau are listed along with the decrease in their area and the time duration (Tables 2.11 and 2.12).

2.4 Conclusion

The temperature of the Earth's environment is increasing in a noticeable pattern as discussed in the chapter above, and this increase is mostly due to the anthropogenic activities. These anthropogenic activities have increased the concentration of the carbon dioxide (CO_2) gas which is the major cause of global warming and the main culprit of climate change. This global warming has given heat to the environmental issues resulted due to the climate change like heat waves, droughts, cyclones, glacier melting furthermore this has also affected the food systems, economic and social conditions of the respective countries as discussed above in the chapter. Now there is an immediate need to overcome and control the increasing carbon dioxide concentration to prevent the impacts and extreme events due to global warming.

References

Anwar MN, Fayyaz A, Sohail NF, Khokhar MF, Baqar M, Khan WD et al (2018) CO2 capture and storage: a way forward for sustainable environment. J Environ Manag 226:131–144. https://doi.org/10.1016/j.jenvman.2018.08.009

Arnone Iii JA, Verburg PS, Johnson DW, Larsen JD, Jasoni RL, Lucchesi AJ et al (2008) Prolonged suppression of ecosystem carbon dioxide uptake after an anomalously warm year. Nature 455 (7211):383

Azevedo VG, Sartori S, Campos LM (2018) CO2 emissions: A quantitative analysis among the BRICS nations. Renew Sust Energ Rev 81:107–115. https://doi.org/10.1016/j.rser.2017.07.027

Bender MA, Knutson TR, Tuleya RE, Sirutis JJ, Vecchi GA, Garner ST, Held IM (2010) Modeled impact of anthropogenic warming on the frequency of intense Atlantic hurricanes. Science 327 (5964):454–458

Boden TA, Marland G, Andres RJ (2017) Global, regional, and national fossil-fuel CO2 emissions. Carbon Dioxide Information Analysis Center, Oak Ridge National Laboratory, U.S. Department of Energy, Oak Ridge. https://doi.org/10.3334/CDIAC/00001_V2017

Burton MR, Sawyer GM, Granieri D (2013) Deep carbon emissions from volcanoes. Rev Mineral Geochem 75(1):323–354. https://doi.org/10.2138/rmg.2013.75.11

Chandel SS, Shrivastva R, Sharma V, Ramasamy P (2016) Overview of the initiatives in renewable energy sector under the national action plan on climate change in India. Renew Sust Energ Rev 54:866–873. https://doi.org/10.1016/j.rser.2015.10.057

Chiodini G (2005) Carbon dioxide diffuse degassing and estimation of heat release from volcanic and hydrothermal systems. J Geophys Res 110(B8)

Ciais P, Reichstein M, Viovy N, Granier A, Ogée J, Allard V et al (2005) Europe-wide reduction in primary productivity caused by the heat and drought in 2003. Nature 437(7058):529. https://doi.org/10.1038/nature03972

Co2 emissions from fuel combustion (2017). Retrieved from www.iea.org

Cullen NJ, Mölg T, Kaser G, Hussein K, Steffen K, Hardy DR (2006) Kilimanjaro Glaciers: recent areal extent from satellite data and new interpretation of observed 20th century retreat rates. Geophys Res Lett 33(16). https://doi.org/10.1029/2006GL027084

da Rocha RP, Reboita MS, Gozzo LF, Dutra LMM, de Jesus EM (2018) Subtropical cyclones over the oceanic basins: a review. Ann N Y Acad Sci 1436:138–156. https://doi.org/10.1111/nyas.13927

Dinan T (2017) Projected increases in hurricane damage in the United States: the role of climate change and coastal development. Ecol Econ 138:186–198

Dolan R, Davis RE (1992) An intensity scale for Atlantic coast northeast storms. J Coast Res:840–853

Duvat VK, Volto N, Salmon C (2017) Impacts of category 5 tropical cyclone Fantala (April 2016) on Farquhar Atoll, Seychelles Islands, Indian Ocean. Geomorphology 298:41–62. https://doi.org/10.1016/j.geomorph.2017.09.02

Easterling W, Apps M (2005) Assessing the consequences of climate change for food and forest resources: a view from the IPCC, In Increasing Climate Variability and Change. Springer, Dordrecht, pp 165–189

Easterling WE, Aggarwal PK, Batima P, Brander KM, Erda L, Howden SM et al (2007) Food, fibre and forest products. Climate Change:273–313

Falloon P, Betts R (2010) Climate impacts on European agriculture and water management in the context of adaptation and mitigation—the importance of an integrated approach. Sci Total Environ 408(23):5667–5687. https://doi.org/10.1016/j.scitotenv.2009.05.002

FAO (2008) The state of food insecurity in the world 2008, high food prices and food security threats and opportunities. Food and Agriculture Organization of the United Nations, Rome, Italy

FAO (2009) The State of Agricultural Commodity Markets 2009. Food and Agriculture Organization of the United Nations, Rome

Francou B, Ramirez E, Cáceres B, Mendoza J (2000) Glacier evolution in the tropical Andes during the last decades of the 20th century: Chacaltaya, Bolivia, and Antizana, Ecuador. AMBIO J Hum Environ 29(7):416–423

Gao H, Li H, Duan Z, Ren Z, Meng X, Pan X (2018) Modelling glacier variation and its impact on water resource in the Urumqi Glacier No. 1 in Central Asia. Sci Total Environ 644:1160–1170

García-Llamas P, Suárez-Seoane S, Taboada A, Fernández-Manso A, Quintano C, Fernández-García V et al (2019) Environmental drivers of fire severity in extreme fire events that affect Mediterranean pine forest ecosystems. For Ecol Manag 433:24–32. https://doi.org/10.1016/j.foreco.2018.10.051

Garnier J, Le Noë J, Marescaux A, Sanz-Cobena A, Lassaletta L, Silvestre M et al (2019) Long-term changes in greenhouse gas emissions from French agriculture and livestock (1852–2014): from traditional agriculture to conventional intensive systems. Sci Total Environ 660:1486–1501. https://doi.org/10.1016/j.scitotenv.2019.01.048

Ghommem M, Hajj MR, Puri IK (2012) Influence of natural and anthropogenic carbon dioxide sequestration on global warming. Ecol Model 235:1–7

Gillett NP, Stott PA, Santer BD (2008) Attribution of cyclogenesis region sea surface temperature change to anthropogenic influence. Geophys Res Lett 35(9). https://doi.org/10.1029/2008GL033670

Haeberli W, Hohmann R (2008) Climate, glaciers and permafrost in the Swiss Alps 2050: scenarios, consequences and recommendations. In: Proceedings ninth international conference on Permafrost, vol 1. Institute of Northern Engineering, University of Alaska Fairbanks, pp 607–612

Handmer J, Honda Y, Kundzewicz ZW, Arnell N, Benito G, Hatfield J, Mohamed IF, Peduzzi P, Wu S, Sherstyukov B, Takahashi K, Yan Z (2012) Changes in impacts of climate extremes: human systems and ecosystems. In: Field CB, Barros V, Stocker TF, Qin D, Dokken DJ, Ebi KL, Mastrandrea MD, Mach KJ, Plattner G-K, Allen SK, Tignor M, Midgley PM (eds) Managing the risks of extreme events and disasters to advance climate change adaptation. A Special Report of Working Groups I and II of the Intergovernmental Panel on Climate Change (IPCC). Cambridge University Press, Cambridge, UK/New York, NY, USA, pp 231–290

Heim RR Jr (2002) A review of twentieth-century drought indices used in the United States. Bull Am Meteorol Soc 83(8):1149–1166

Ho DTK (2018) Climate change in Malaysia: Trends, contributors, impacts, mitigation and adaptations. Sci Total Environ

Huey RB, Deutsch CA, Tewksbury JJ, Vitt LJ, Hertz PE, Álvarez Pérez HJ, Garland T Jr (2009) Why tropical forest lizards are vulnerable to climate warming. Proc R Soc B Biol Sci 276 (1664):1939–1948

Im ES, Pal JS, Eltahir EA (2017) Deadly heat waves projected in the densely populated agricultural regions of South Asia. Sci Adv 3(8):e1603322. https://doi.org/10.1126/sciadv.1603322

IPCC (2014) In: Core Writing Team, Pachauri RK, Meyer LA (eds) Climate change 2014: synthesis report. Contribution of working groups I, II and III to the fifth assessment report of the intergovernmental panel on climate change. IPCC, Geneva, Switzerland. 151 pp

Kamaruddin AH, Din AHM, Pa'suya MF, Omar KM (2016) Long-term sea level trend from tidal data in Malaysia. Paper presented at the Control and System Graduate Research Colloquium (ICSGRC), 2016 7th IEEE

Karoly DJ, Wu Q (2005) Detection of regional surface temperature trends. J Clim 18 (21):4337–4343

Knutson TR, McBride JL, Chan J, Emanuel K, Holland G, Landsea C et al (2010) Tropical cyclones and climate change. Nature Geosci 3(3):157. https://doi.org/10.1038/ngeo779

Kossin JP, Camargo SJ, Sitkowski M (2010) Climate modulation of North Atlantic hurricane tracks. J Clim 23(11):3057–3076. https://doi.org/10.1175/2010JCLI3497.1

LaMarche VC, Graybill DA, Fritts HC, Rose MR (1984) Increasing atmospheric carbon dioxide: tree ring evidence for growth enhancement in natural vegetation. Science 225 (4666):1019–1021. https://doi.org/10.1126/science.225.4666.1019

Le Quéré C, Andres RJ, Boden T, Conway T, Houghton RA, House JI et al (2012) The global carbon budget 1959–2011. Earth Syst Sci Data Discuss 5(2):1107–1157. https://doi.org/10.5194/essdd-5-1107-2012

Leclercq PW, Oerlemans J, Cogley JG (2011) Estimating the glacier contribution to sea-level rise for the period 1800–2005. Surv Geophys 32(4–5):519–535. https://doi.org/10.1007/s10712-011-9121-7

Lee JY, Ellingwood BR (2017) A decision model for intergenerational life-cycle risk assessment of civil infrastructure exposed to hurricanes under climate change. Reliab Eng Syst Saf 159:100–107

Lemke P, Ren J, Alley RB, Allison I, Carrasco J, Flato G et al (2007) Observations: changes in snow, ice and frozen ground, climate change 2007: The physical science basis. Contribution of working group i to the fourth assessment report of the Intergovernmental Panel on Climate Change. Cambridge University Press, Cambridge, pp 337–383

Letcher TM (ed) (2018) Managing Global Warming: an interface of technology and human issues. Academic. https://doi.org/10.1016/B978-0-12-814104-5.00001-6

Li X, Kang S, Zhang G, Qu B, Tripathee L, Paudyal R et al (2018) Light-absorbing impurities in a southern Tibetan plateau glacier: variations and potential impact on snow albedo and radiative forcing. Atmos Res 200:77–87

Liggins F, Betts RA, McGuire B (2010) Projected future climate changes in the context of geological and geomorphological hazards. Philos Trans R Soc A Math Phys Eng Sci 368 (1919):2347–2367. https://doi.org/10.1098/rsta.2010.0072

Lobell DB, Schlenker W, Costa-Roberts J (2011) Climate trends and global crop production since 1980. Science 333(6042):616–620. https://doi.org/10.1126/science.1204531

Luthi D, Le Floch M, Bereiter B, Blunier T, Barnola J-M, Siegenthaler U, Raynaud D, Jouzel J, Fischer H, Kawamura K, Stocker TF (2008) High-resolution carbon dioxide concentration record 650,000-800,000 years before present. Nature 453(7193):379–382

Mardones C, Flores B (2018) Effectiveness of a CO2 tax on industrial emissions. Energy Econ 71:370–382. https://doi.org/10.1016/j.eneco.2018.03.018

McGuire B (2010) Potential for a hazardous geospheric response to projected future climate changes. Philos Trans R Soc A Math Phys Eng Sci 368(1919):2317–2345. https://doi.org/10.1098/rsta.2010.0080

Meehl GA, Tebaldi C (2004) More intense, more frequent, and longer lasting heat waves in the 21st century. Science 305(5686):994–997. https://doi.org/10.1126/science.1098704

Müller C, Cramer W, Hare WL, Lotze-Campen H (2011) Climate change risks for African agriculture. Proc Natl Acad Sci 108(11):4313–4315

Nejat P, Jomehzadeh F, Taheri MM, Gohari M, Majid MZA (2015) A global review of energy consumption, CO2 emissions and policy in the residential sector (with an overview of the top ten CO2 emitting countries). Renew Sust Energ Rev 43:843–862. https://doi.org/10.1016/j.rser.2014.11.066

OECD-FAO (2008) OECD-FAO agricultural outlook 2008–2017, Highlights. Organization for Economic Co-operation and Development and Food and Agriculture Organization, Paris, France. www.fao.org/es/ESC/common/ecg/550/en/AgOut2017E.pdf

Pachauri RK, Allen MR, Barros VR, Broome J, Cramer W, Christ R, et al (2014) Climate change 2014: synthesis report. Contribution of Working Groups I, II and III to the fifth assessment report of the Intergovernmental Panel on Climate Change: IPCC

Pall P, Aina T, Stone DA, Stott PA, Nozawa T, Hilberts AG et al (2011) Anthropogenic greenhouse gas contribution to flood risk in England and Wales in autumn 2000. Nature 470 (7334):382–385. https://doi.org/10.1038/nature09762

Pappas D, Chalvatzis KJ, Guan D, Ioannidis A (2018) Energy and carbon intensity: A study on the cross-country industrial shift from China to India and SE Asia. Appl Energy 225:183–194. https://doi.org/10.1016/j.apenergy.2018.04.132

Paul F, Kääb A, Maisch M, Kellenberger T, Haeberli W (2004) Rapid disintegration of Alpine glaciers observed with satellite data. Geophys Res Lett 31(21). https://doi.org/10.1029/2004GL020816

Peters O, Hertlein C, Christensen K (2001) A complexity view of rainfall. Phys Rev Lett 88 (1):018701. https://doi.org/10.1103/PhysRevLett.88.018701

Potter C (2018) Ecosystem carbon emissions from 2015 forest fires in interior Alaska. Carbon Balance Manag 13(1):2. https://doi.org/10.1186/s13021-017-0090-0

Rahman S, Chang HC, Magill C, Tomkins K, Hehir W (2018) Forest fire occurrence and modeling in Southeastern Australia. In Forest Fire IntechOpen. https://doi.org/10.5772/intechopen.76072

Reboita M, da Rocha R, Oliveira D (2019) Key features and adverse weather of the named subtropical cyclones over the Southwestern South Atlantic Ocean. Atmos 10(1):6. https://doi.org/10.3390/atmos10010006

Reichert BK, Bengtsson L, Oerlemans J (2002) Recent glacier retreat exceeds internal variability. J Clim 15(21):3069–3081

Ritchie H, Roser M (2017) Co2 and other greenhouse gas emissions. Retrieved from https://ourworldindata.org/co2-and-other-greenhouse-gas-emissions

Russo S, Dosio A, Graversen RG, Sillmann J, Carrao H, Dunbar MB et al (2014) Magnitude of extreme heat waves in present climate and their projection in a warming world. J Geophys Res Atmos 119(22):12–500. https://doi.org/10.1002/2014JD022098

Sarker MA (2018) Numerical modelling of waves and surge from Cyclone Chapala (2015) in the Arabian Sea. Ocean Eng 158:299–310

Shakerian F, Kim K-H, Szulejko JE, Park J-W (2015) A comparative review between amines and ammonia as sorptive media for post-combustion CO 2 capture. Appl Energy 148:10–22

Shi A (2003) The impact of population pressure on global carbon dioxide emissions, 1975–1996: evidence from pooled cross-country data. Ecol Econ 44(1):29–42. https://doi.org/10.1016/S0921-8009(02)00223-9

Smith LT, Aragao LE, Sabel CE, Nakaya T (2014) Drought impacts on children's respiratory health in the Brazilian Amazon. Sci Rep 4:3726. https://doi.org/10.1038/srep03726

Stott PA, Stone DA, Allen MR (2004) Human contribution to the European heatwave of 2003. Nature 432(7017):610. https://doi.org/10.1038/nature03130

Tans P (2008) Trends in atmospheric carbon dioxide: Mauna Loa. NOAA Earth System Research Laboratory (ESRL) 1959–2008 data

Victor DG, Zhou D, Ahmed EHM, Dadhich PK, Olivier JGJ, Rogner H-H, Sheikho K, Yamaguchi M (2014) Introductory chapter. In: Edenhofer O, Pichs-Madruga R, Sokona Y, Farahani E, Kadner S, Seyboth K, Adler A, Baum I, Brunner S, Eickemeier P, Kriemann B, Savolainen J, Schlömer S, von Stechow C, Zwickel T, Minx JC (eds) Climate Change 2014: Mitigation of Climate Change. Contribution of working group III to the fifth assessment report of the Intergovernmental Panel on Climate Change. Cambridge University Press, Cambridge/New York

Viste E, Korecha D, Sorteberg A (2013) Recent drought and precipitation tendencies in Ethiopia. Theor Appl Climatol 112(3–4):535–551. https://doi.org/10.1007/s00704-012-0746-3

Waheed R, Chang D, Sarwar S, Chen W (2017) Forest, agriculture, renewable energy, and CO2 emission. J Clean Prod 30:1e8. https://doi.org/10.1016/j.jclepro.2017.10.287

Wang Q, Zhao M, Li R, Su M (2018) Decomposition and decoupling analysis of carbon emissions from economic growth: a comparative study of China and the United States. J Clean Prod 197:178–184. https://doi.org/10.1016/j.jclepro.2018.05.285

Webster PJ, Holland GJ, Curry JA, Chang HR (2005) Changes in tropical cyclone number, duration, and intensity in a warming environment. Science 309(5742):1844–1846. https://doi.org/10.1126/science.1116448

Yang JP, Ding YJ, Liu SY, Tan CP (2015) Vulnerability of mountain glaciers in China to Climate Change. Adv Clim Chang Res 6(3–4):171–180

Zemp M, Hoelzle M, Haeberli W (2009) Six decades of glacier mass-balance observations: a review of the worldwide monitoring network. Ann Glaciol 50(50):101–111

Chapter 3
Carbon Capture Utilization and Storage Supply Chain: Analysis, Modeling and Optimization

Grazia Leonzio

Abstract In the 2016, CO_2 emissions in the world was about 32.3 Gt with the combustion of fossil fuel being the highest contributor. A target for the reduction of CO_2 emissions of 60% was set in the Paris Agreement. Thereafter, this topic has acquired much importance in the world, especially in recent years. In this context, carbon capture utilization and its storage supply chain is highly considered as a strategic solution, that can solve the problems related to CO_2 emissions. In this work, an overview about carbon capture utilization and its storage supply chain is developed. Due to their important environmental role, mathematical models are required for design and optimization. Hierarchical or simultaneous methodology can be used, and one procedure is suggested for minimizing the total cost of supply chain. Following this, equations related to capture and compression costs, transportation costs, utilization and storage costs are revised. This work, in addition, reviews systems for CO_2 capture and compression, and options for CO_2 utilization and storage. Many work are present in literature regarding this technology, however more studies should be developed on dynamic state considering uncertainties.

Keywords Carbon capture utilization and storage supply chain · Mathematical modeling · Optimization · Design · CO_2 capture technology · CO_2 utilization · CO_2 storage · CO_2 reduction · Economic analysis · Uncertainties

3.1 Introduction

The variation of climate has a high influence on all aspects of society and it is recognized as public concern in the United Nations Frame-work Convention on Climate Change (UNFCCC) (Pires et al. 2011; Han and Lee 2011). In this contest, carbon capture, utilization and storage (CCUS) systems have an important role and researchers worldwide have focused their attention on carbon capture utilization and storage supply chains. Figure 3.1 shows the main elements of carbon capture

G. Leonzio (✉)
Department of Industrial and Information Engineering and Economics, University of L'Aquila, L'Aquila, Italy
e-mail: grazia.leonzio@graduate.univaq.it

Inamuddin et al. (eds.), *Sustainable Agriculture Reviews 37*, Sustainable Agriculture Reviews 37, https://doi.org/10.1007/978-3-030-29298-0_3

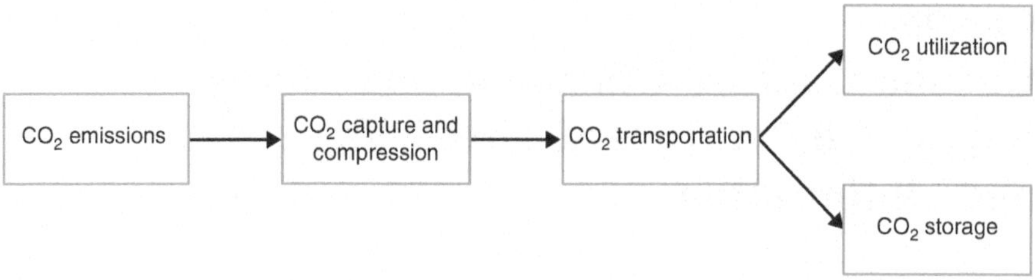

Fig. 3.1 Block diagram for the main elements (CO_2 source, CO_2 capture and compression, CO_2 transportation, CO_2 utilization and CO_2 storage) of carbon capture utilization and storage supply chain (Hasan et al. 2015)

utilization and storage supply chain: CO_2 source, CO_2 capture technology/material and compression, CO_2 transportation, CO_2 storage and CO_2 utilization (Hasan et al. 2015).

When more than one CO_2 sources, capture technologies/materials, CO_2 utilization and storage sites are present, the system is known as carbon capture utilization and storage supply chain network and the best combination between single elements then the best supply chain should be found.

There is international interest to carbon capture utilization and storage systems, because inside the sustainable economy they can limit the increase of temperature to 2 °C, as established by the Paris Agreement (COP21) setting a reduction target of 50% for CO_2 emissions by 2050 (Tapia et al. 2018). The Intergovernmental Panel on Climate Change (IPCC) suggests that the cost of actions to solve this environmental problem should increase by 138% without carbon capture utilization and storage systems and the certainty to achieve the fixed objective is not present. In particular, climate action without CCUS systems are calculated around $2 trillion over 40 years. The application of carbon capture utilization and storage systems in industrial and power sectors should reduce greenhouse gas emissions of 7 Gigatonnes per year by 2050 with the aim to constrain the increase of heating of 2 °C. The importance of carbon capture utilization and storage system can be understood by looking at emission data: despite traditional climate change mitigation initiatives and polices, global greenhouse gas emissions grew by 2.2% every year between 2000 and 2010 compared to the increase of 1.3% every year between 1970 and 2000 (Milani et al. 2015). Therefore, in order to reach the objectives set by the Paris Climate Change Agreement, carbon capture utilization and storage will be an important technology, especially for the long term and will have a promising value (Sun and Chen 2017).

Many mathematical models are performed with the aim to design and optimize carbon capture utilization and storage supply chain by minimizing the total costs or by maximizing the amount of captured CO_2. The aim is to find, inside a carbon capture utilization and storage supply chain network, the best combination between CO_2 source, CO_2 capture and compression, CO_2 transportation, CO_2 storage and CO_2 utilization. CO_2 sources and CO_2 utilization/storage sites are connected as in a combinatorial problem: more alternatives are present by increasing CO_2 sources and utilization/storage sites. An overview of CO_2 capture technology, CO_2 storage and

CO_2 utilization is suggested in this work and its aim is to suggest a guide to model carbon capture utilization and storage supply chain. Then a review of work presented in literature is shown.

3.2 Status of Carbon Capture Utilization and Storage Supply Chain

Overall, in the world the realization of carbon capture utilization and storage supply chains is slower than the objectives established by Paris Agreement cannot be satisfied.

70 carbon capture utilization and storage projects at pilot scale are present in the world: 22 are in North America, 1 in South America, 22 in Europe, 20 in Asia, 4 in Australia, and 1 in South Africa, as shown in Fig. 3.2 (Liu et al. 2017).

Some important projects related to carbon capture utilization and storage supply chains are: Petra Nova in Texas (CO_2 emitted by coal-fired power plant is used to enhanced oil recovery), Al Reyadah in Abu Dhabi (CO_2 from iron and steel industry is used to enhanced oil recovery), Tuticorin in India (CO_2 emitted by coal powered boiler is used to make baking soda), Boundary Dam 3 in Canada (CO_2 from power plant is used to enhanced oil recovery), Air Products in Texas (CO_2 from hydrogen

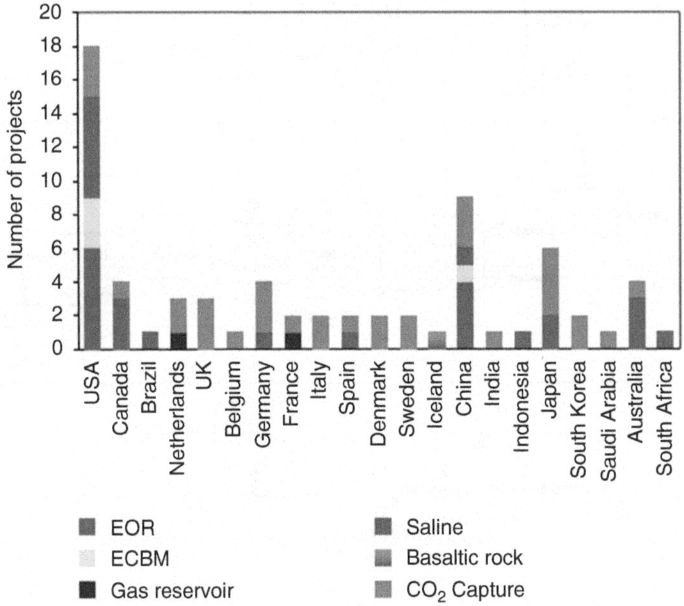

Fig. 3.2 Global distribution of pilot-scale carbon capture utilization and storage engineering projects based on project purpose and reservoir types (http://www.globalccsinstitute.com/) (*EOR* enhanced oil recovery, *ECBM* enhance coal-bed methane)

refining is used to enhanced oil recovery), Emirates Steel Project in UAE (CO_2 from steel industry is utilized to enhanced oil recovery), Yangchang Petroleum in China (CO_2 from coal industry is used to enhanced oil recovery). Other small- or large-scale carbon capture utilization and storage projects have been developed showing significative reductions of CO_2. However, many studies and research has been carried out to estimate the costs of these supply chains, the main obstacle for their development at industrial scale.

3.3 Carbon Capture Utilization and Storage Technology Overview

3.3.1 CO_2 Capture Options

CO_2 can be captured from flue gases with three strategies: post-combustion, pre-combustion, oxy-combustion, as present in Fig. 3.3.

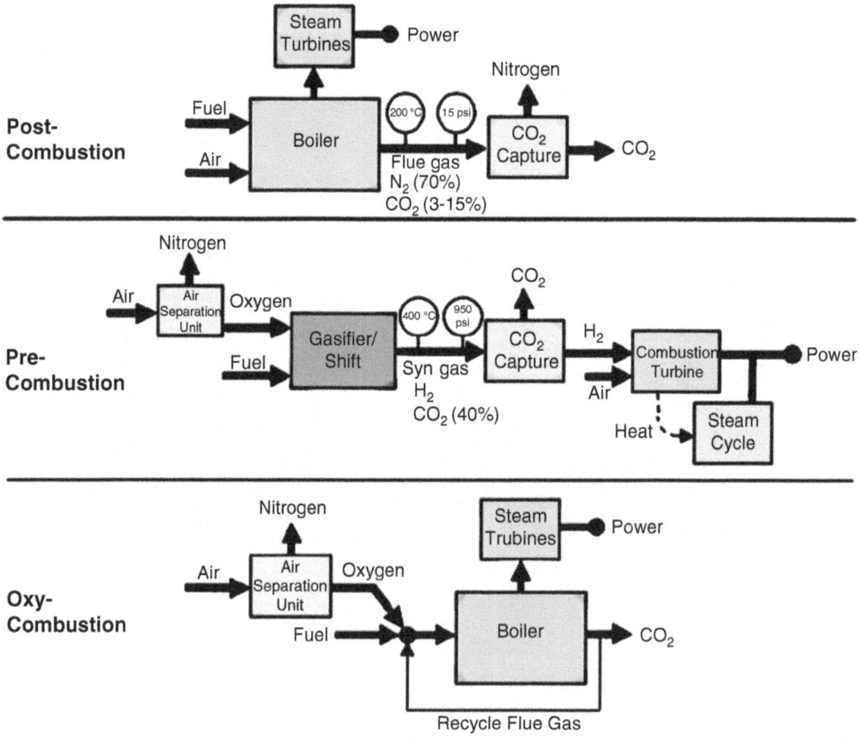

Fig. 3.3 Block diagram for different CO_2 capture strategies (post-combustion, pre-combustion and oxy-combustion) (Figueroa et al. 2008, reproduced with permission number 4563270013378)

In post-combustion capture, CO_2 is captured from flue gas obtained by the burning with air. In pre-combustion capture, CO_2 is removed by syngas obtained by gasification process, before the combustion. In oxy-combustion, unlike to post-combustion, heat source is combusted with oxygen containing little or no nitrogen and flue gas is recycled after that CO_2 is captured. The selection of one of these strategies depends on advantages and disadvantages, CO_2 concentration and pressure. Advantages for post-combustion capture are: a straightforward approach to be retrofitted and it is a more mature strategy respect to other strategies so that can be more applied to existing power plant. However, some disadvantages are: a low CO_2 concentration (about 5–15 vol%) that due to the near atmospheric pressure does not cause high CO_2 partial pressure (about lower than 0.15 atm), the requirement of higher circulation volume for high capture level producing energy lost during solvent/sorbent regeneration (Figueroa et al. 2008; Song et al. 2019). Advantages for pre-combustion are: a high CO_2 concentration (about 45 vol%) and pressure, then a high CO_2 partial pressure and driving force for the separation. For these reasons, a lot of systems can be used for CO_2 capture, however severe operating conditions are required (15–20 bar and 190–210 °C) and as the previous strategy energy penalties are present due to sorbent regeneration (Figueroa et al. 2008; Song et al. 2019; Dai et al. 2016; Nandi et al. 2015). Advantages for oxy-combustion are: a low investment cost caused by a high CO_2 concentration (80–98 vol%); however, efficiencies are not so high determining energy penalties, the production of O_2 can be very expensive and auxiliary load are added (Figueroa et al. 2008; Song et al. 2019; Hedin et al. 2013; Tonziello and Vellini 2011; Leung et al. 2014). Generally, post-combustion strategies are used for power generators fueled by coal and air. Pre-combustion strategies are used for gasification process. Oxy-combustion strategy is generally used for already existing plants. Different technologies as absorption, adsorption, membrane, chemical looping and cryogenic can be used to capture CO_2 in the above strategies. These are based on different principles (Table 3.1).

3.3.1.1 Absorption Technology

Absorption process can be physical or chemical (Song et al. 2019). Absorber and regenerator work in a continuous way: flue gas (fed at the bottom of the absorber) and solvent (fed at the top of the absorber) are working in counter current way then a selective capture of CO_2 is present. Stream with a high CO_2 content from the absorber is sent to the regenerator: CO_2 is desorbed and it can be used or stored, while the solvent is regenerated and recycled for further use in the absorber. Chemical solution for absorption such as monoethanol amine (MEA), diethanol amine (DEA), Nmethyldiethanolamine (MDEA), and di-2-propanolamine (DIPA) are generally used (Mamun et al. 2007). Diglycolamine (DGA), 2-(2-aminoethylamino) ethanol (AEE), 2-amino 2-methyl 1-propanol (AMP), N-2-aminoethyl 1,3-propanediamine (AEPDNH2), triethanol amine (TEA), triethylene tetra amine (TETA), piperazine (PZ), glucosamine (GA), NaOH, NH_3, K_2CO_3, KOH, Na_2CO_3, etc. are other absorbent solutions that can be used, but with some

Table 3.1 Suggested CO_2 capture technologies for different CO_2 capture strategies with the description of methodology and applications (Cuellar-Franca and Azapagic 2015)

Capture option	Separation technology	Method	Applications
Post-conversion	Absorption by chemical solvent	Amine-based solvent, e.g. monoethanolamine (MEA), diethanolamine (DEA), and hindered amine (KS-1)	Power plants; iron and steel industry; cement industry; oil refineries
		Alkaline solvents, e.g. NaOH and Ca $(OH)_2$	
		Ionic liquids	
	Adsorption by solid sorbents	Amine-based solid sorbents	No application reported
		Alkali earth metal-based solid sorbents, e.g. $CaCO_3$	
		Alkali metal carbonate solid sorbents, e.g. Na_2CO_3 and K_2CO_3	
		Porous organic frameworks – Polymers	Power plants
	Membrane separation	Polymeric membranes, e.g. polymeric gas permeation membranes	Power plants; natural gas sweetening
		Inorganic membranes, e.g. zeolites	
		Hybrid membranes	
	Cryogenic separation	Cryogenic separation	Power plants
	Pressure/vacuum swing adsorption	Zeolites	Power plants; iron and steel industry
		Activated carbon	
Pre-conversion	Absorption by physical solvents	Selexol, rectisol	Power plants (IGCC)
	Absorption by chemical solvents	Amine-based solvent, e.g. Monoethanolamine (MEA)	Ammonia production
	Adsorption by porous organic framework	Porous organic frameworks membranes	Gas separations
Oxy-fuel combustion	Separation of oxygen from air	Oxy-fuel process	Power plants; iron and steel industry; cement industry
		Chemical looping combustion	Power plants
		Chemical looping reforming	Power plants; syngas production and upgrading

limitations for application in large scale plants (degradation, corrosion, regeneration efficiency). In recent years, ionic liquids are under research for application in CO_2 absorption process. Due to their physical-chemical properties (low volatility, good dissolution, high decomposition temperature and stability, excellent chemical tunabilities), they can reduce regeneration energy but until now they have a higher cost compared to other solvents (Luo and Wang 2017; Ma et al. 2018). Some advantages of absorption process are: chemical solvents ensure a high driving force to capture CO_2 also at low concentration, while wet-scrubbing guarantees a recovery of heat especially for exothermic absorption reactions (National Energy Technology Laboratory 2010b). Absorption technology is less expensive for large-scale plants, simpler and more robust than other technologies. For these reasons, they are preferred and applied among capture technologies (Bhown 2014). Actually, researches in absorption system for CO_2 capture are regarding the improvements of solvents and gas-liquid contactor, the study of inhibitors for degradation and corrosion, the study of alternative solvents and blended solvents, etc.

3.3.1.2 Adsorption Technology

In the first section of this process, flue gas is dried for the condensation of water and after compression it is sent to chambers packed with solid adsorbent (activated carbon, zeolites, or metal organic frameworks, etc.) (Gao et al. 2017; Samanta et al. 2012). Generally, in one system, two or three adsorption chambers are present: the first receives the feed for CO_2 adsorption, the second for its desorption, while the last is in stand-by for the first one (Thiruvenkatachari et al. 2009). The system operates continuously, changing pressure in PSA (pressure swing adsorption) and VPSA (vacuum pressure swing adsorption) systems, temperature in TSA (temperature swing adsorption) systems or voltage electric current in electrical swing adsorption (ESA) systems. It is possible to combine pressure swing adsorption and temperature swing adsorption in PTSA systems (pressure and temperature swing adsorption) that could reduce power consumption by 11% respect to pressure swing adsorption system (Gupta et al. 2003). However, adsorption is lower used for large scale processes than absorption process (Bhown 2014). In any cases, some advantages for adsorption process are: fast kinetic, the large adsorption capacity of sites and low sensible heat, it allows to work with low CO_2 concentration and at a higher capacities than absorption (National Energy Technology Laboratory 2010b). Some disadvantages for adsorption are related to the regeneration and reusability of adsorbents and a lower efficiency and selectivity respect to absorption or cryogenic technology (Bamdad et al. 2018; Mondal et al. 2012). Researches in adsorption process are about the study of novel adsorbents or the modification of already present (their surface) with the aim to work at high temperatures and with steam, resulting in a high capacity and selectivity. In this context, a good solution is presented by activated carbon fibers and carbon fiber composites.

3.3.1.3 Membrane Technology

Membrane technology is introduced in last years and it is divided in gas separation and gas absorption. In the suggested process the permeate of the first stage is divided in two parts: 70% is fed to the second-stage while 30% is recycled to the first stage as sweep (Hussain and Hägg 2010). Instead, the permeate of the second stage is recycled to the same stage as a sweep for only 5%. These systems are a new technologies and have some disadvantages related to higher temperature than 100 °C, sensibility to corrosion due to particular gases, high performance for long term operation, the requirement of multiple stages for diluted feed stream (below 20%) and inability to handle high flue gas flow rate (Zhang et al. 2013a, b; Mondal et al. 2012). Some advantages are due to the no use of steam, chemicals and its simple design (National Energy Technology Laboratory 2010b). Also, membrane systems are competitive due its energy efficiency and sustainability (Evangelos et al. 2017). Generally, polymer membranes and composite membranes are used in addition to inorganic one (carbon, zeolite, ceramic or metallic) and can be porous and not. In last years, mixed matrix membranes are developed and are composed by a polymer with dispersed zeolites inside. This solution can improve the selectivity of membrane compared to traditional one (Li et al. 2013a, b). It is evident, that today researches are about the finding new efficient materials.

3.3.1.4 Chemical Looping Combustion

In a chemical looping combustion (CLC) for carbon capture system, combustion is divided in two steps: oxidation and reduction reaction by using oxygen (in the form of solid metal oxides as Fe_2O_3, NiO, CuO, Mn_2O_3 called solid oxygen carrier), moving between the two separated phases, with two fluidized bed reactors, one for air and one for fuel (Song et al. 2019). In the first reactor, metal oxides are oxidized by the oxygen of air. In the second reactor, metal oxides are reduced by fuel, which is oxidized to CO_2 and H_2O. The generalized reaction for fuel reactor is the following (Mondal et al. 2012) (see Eq. 3.1):

$$(2n + m)M_yO_x + C_nH_{2m} \rightarrow (2n + m)M_yO_{x-1} + mH_2O + nCO_2 \tag{3.1}$$

while for air reactor, the following reaction take places (Mondal et al. 2012) (see Eq. 3.2)

$$M_yO_{x-1} + \frac{1}{2}O_2(air) \rightarrow M_yO_x + N_2 + excessO_2 \tag{3.2}$$

Advantages of this technology are: the production of no toxic N_2 from air reactor and the production of CO_2 from fuel reactor that can be separated by H_2O by condensation then reducing capital costs (Olajire 2010). In addition, NO_x formations are minimized, because combustion is carried out with air reducing oxygen that is re-oxidized in the air reactor at lower temperature. However, there are few large-scale plants using this process (most of them are in laboratory scale) with some problems, as the low stability of oxygen carrier, the slow reaction rate of redox reaction and the removal of sulfur by fuel to avoid poisoning problems (Solunke and Veser 2011). Chemical looping is especially used in gasification process capturing at least 90% of CO_2: however, the cost of electricity is increased of 16% respect to air fired circulating fluidized bed plant (Nsakala and Liljedahl 2003).

3.3.1.5 Cryogenic Technology

Cryogenic technologies are also known as low temperature CO_2 capture technologies. CO_2 is separated by flue gas due to the different condensation and desublimation properties. High CO_2 recovery (99.99%) and purity (99.99%) can be obtained with this technology (Brunetti et al. 2010). In the process scheme for cryogenic capture a steel monolith structure is used as packed material while liquefied natural gas is used for refrigeration. Energy consumption and installation investment costs are lower than vacuum pressure swing adsorption due to a smaller bed size and even if it has more potential than absorption it is not yet marketable, due to some problems as the losses of sensible and latent heat due to the no good thermal insulation. In addition, in order to remove H_2S, the temperature of system should be about 150 °C, so increasing operating costs. Also, an additional refrigerator is required for liquified natural gas, increasing then the energy consumption of the system (Abatzoglou and Boivin 2009). Other limitations of cryogenic technology are related to the presence of other gases (SO_x, NO_x, H_2O), that during condensation can determine corrosion, fouling phenomena (AxeL and Xiaoshan 1997). Also, CO_2 can produce solids reducing heat transfer and then the efficiency. However, the main advantages of cryogenic technology are that chemical solvents are not necessary and that atmospheric pressure can be set. Also, liquid CO_2 is obtained and it is needed for economical transport in ship and pipeline. In addition to packed bed scheme, other processes are developed with cryogenic technology: anti-sublimation CO_2 capture process (AnSU), CryoCell process, cryogenic distillation and stirling cooler system ensuring a CO_2 recovery between 85% an 99% (Song et al. 2019). Advantages and limitations of these different schemes are reported in Table 3.2. CO_2 can be captured without organic solvents: it is then a green process and it is competitive with refrigerant at low costs. Cryogenic can be used in cases with high CO_2 partial pressure, that it is typical of pre-combustion or oxyfuel combustion process (Gupta et al. 2003). Then, it is not economically favorable for dilute CO_2 stream due the high amount of required energy.

3.3.1.6 Hybrid Technology

With the aim to overcame the limitations of each technology, hybrid capture technology are developed by the combination of different technology as shown in

Table 3.2 Convenience and limitations of cryogenic capture technology (packed bed, AnSU, CryoCell, distillation, stirling cooler)

Category	Cold energy source	Advantages	Limitations
Packed bed	Liquid nitrogen gas (LNG)	Atmosphere	Depends on the availability of LNG
		Simultaneous H_2O and CO_2 removal	Lab scale
		Avoiding high pressure drop	
		Surface area-to-volume ratio of the column	
AnSU	Liquefied natural gas (LNG)	Atmosphere	Depends on the location of natural gas station
		Lower energy penalty than MEA absorption	No H_2O can be tolerated
		Pilot demonstration	Frost CO_2 adversely affects heat conduction undesired mechanical stresses
CryoCell	Chiller	No process heating system required	More suitable for high CO_2 concentration (higher than 20%)
		No corrosion potential	High compression power requirement
		No foaming potential	
		Avoid compression cost	
Distillation	Compressor and cooler	Avoid compression cost	Capital cost for pressure difference
		Easy to be pumped to storage site	High installation cost
		Energy storage potential	
		Water saving potential	
		Simultaneous removal of other pollutants (Hg, SOx, NO_2, HCl, etc.)	
Stirling cooler	Striling cooler	Atmosphere	Exergy loss due to temperature difference
		Simultaneous H_2O and CO_2 removal	Difficulty of frost layer scrapping
		Lower energy penalty than MEA absorption	Lab scale
		Energy storage potential	

Fig. 3.4. Hybrid processes ensure a higher CO_2 recovery and lower energy penalties and installation investment costs. The combination of membrane with absorption process (membrane contactor), catalysis process (adsorption-catalysis-membrane) and cryogenic process (low-temperature-membrane-cryogenic) shows significative

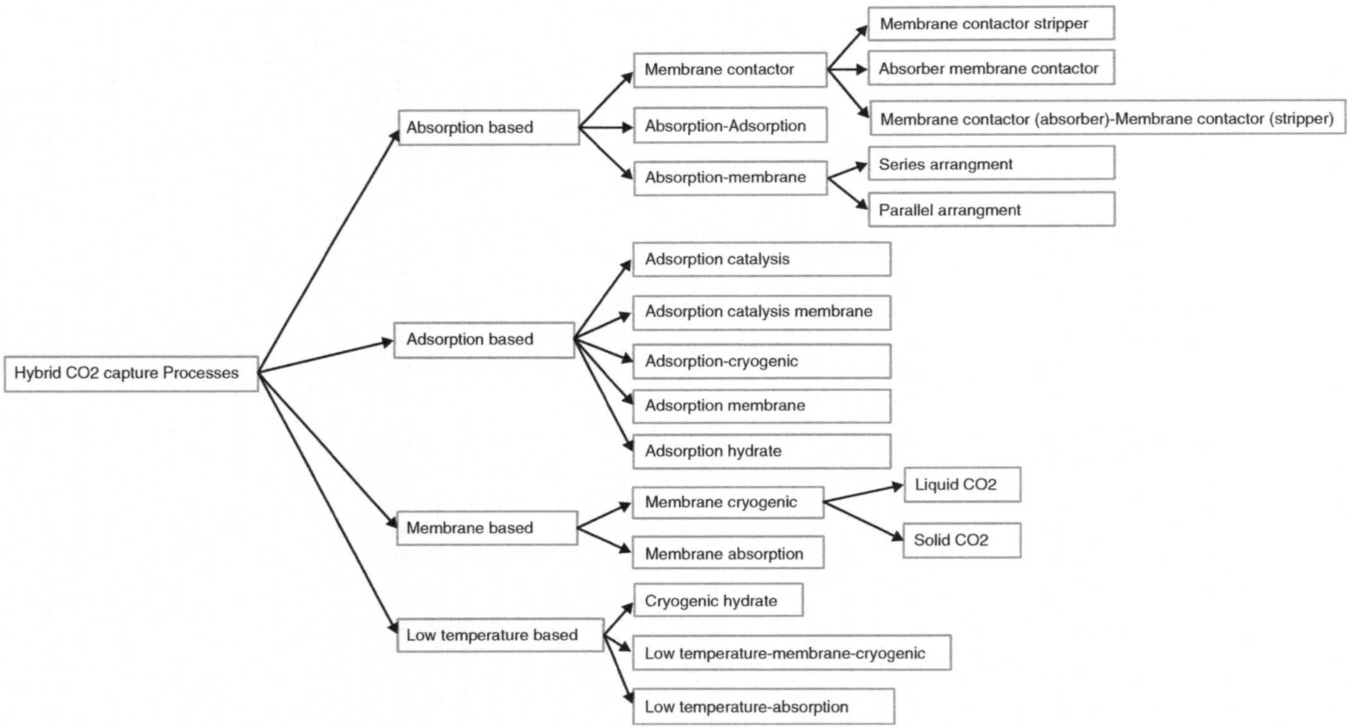

Fig. 3.4 Existing hybrid CO_2 capture processes: absorption based, adsorption based, membrane based and low temperature based (Song et al. 2018, reproduced with permission number 4563280736814)

development. These hybrid technologies can be combined in serie, parallel or integrated. Therefore, hybrid processes is a hopeful alternative for the future.

3.3.2 CO_2 Utilization Options

In addition to storage, CO_2 can be used and then valorized for:

- direct utilization;
- enhanced oil recovery, coal-bed methane recovery and similar, then CO_2 is used as injection fluid;
- conversion into valuable products;
- mineral carbonation;

CO_2 with a high purity is directly used in food and drink industry for carbonation, it is used in preservative production, as packing gas and solvent for flavors extraction and in decaffeination process. Also, CO_2 can be used in medical application or for drugs production (Cuellar-Franca and Azapagic 2015).

In enhanced oil recovery (EOR) and enhance coal-bed methane (ECBM) CO_2 is used to take crude oil from an oil field or natural gas from coal deposit respectively. Enhance coal-bed methane actually is not yet developed, while EOR is widely used in USA in addition to recovery gas from natural reservoirs (Metz et al. 2005). CO_2 can be used also in enhanced shale gas recovery (ESGR), enhanced gas recovery (EGR) and enhanced material recovery (EMR). In this contest, CO_2 is used as injection agent due to a low cost and availability.

CO_2 can be used for conversion into chemicals and fuels, as in Fig. 3.5 where a detailed roadmap is suggested (Ampelli et al. 2015). It is possible to see how many products can be obtained as urea, polymers (polyurethane), methanol, syngas, dimethyl ether, algae, etc. even if actually most of them are at demonstration level and only in the future are commercially available.

A detailed analysis regarding possible fuels that could be produced from CO_2 is shown in Fig. 3.6: formic acid, ethanol, methanol, methane, dimethyl ether, syngas, hydrocarbons and hydrogen are obtained with the respective reactions.

In mineral carbonation, CO_2 reacts with a metal oxide such as magnesium or calcium to produce carbonates. For example, steel slugs obtained by the extraction of alumina from bauxite are sources of metal oxides. However, actually, this route is not used at large scale and high costs are required. Related to mineralization, is the production of concrete, in particular concrete curing and concrete by red mud using CO_2 (Patricio et al. 2017).

A summary of different CO_2 utilization options is shown in Fig. 3.7, considering for each option potential application, economic aspects, energetic consumptions, the amount of required CO_2, the time of sequestration and environmental impact.

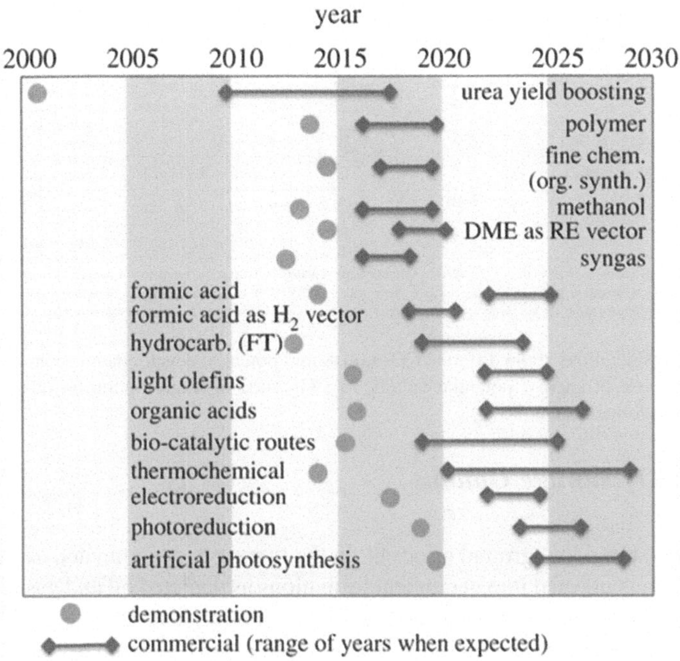

Fig. 3.5 Road map for the commercialization of CO_2 utilization, showing if it is at demonstration or commercial scale (Ampelli et al. 2015)

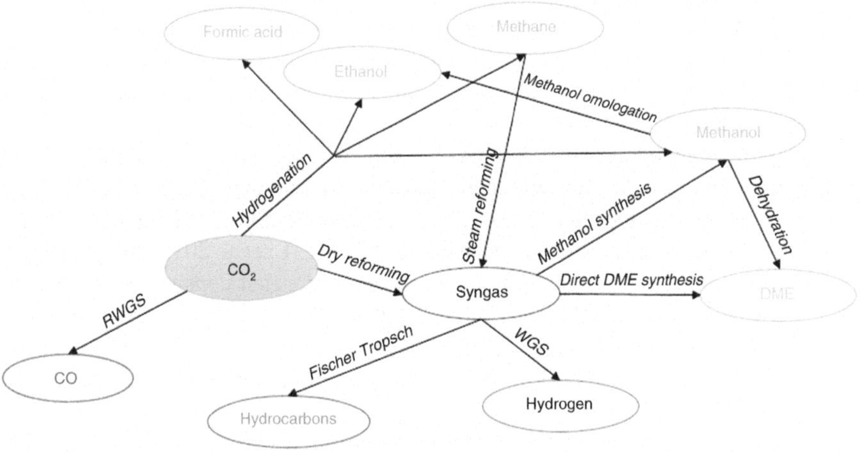

Fig. 3.6 Different routes for CO_2 utilization: products that can be produced by CO_2 and respective reactions (Leonzio 2018)

	already industrial			short term		medium term		long term				
	EOR	industrial use	organic synthesis	hydrogenation	algac-open p.	reforming HC	algac-reactor	mineralization	thermo chem.	electro reduct.	photoelectrochem.	bio catal.
potential development	4	4	4	3	3	2	2	1	1	1	1	1
economic perspectives	4	4	3	3	3	not known	2	1	2	not known	not known	not known
external use of energy	3	3	2	2	4	1	4	1	4	2	4	4
potential vol. of CO_2	2	2	3	4	4	4	4	3	4	4	4	4
time of sequestration	4	1.5	3	2	2	2	2	4	2	2	2	2
other impacts on environment	4	4	3	3	4	1	4	3	3	2	2	3

potential development: 1. more than 10 years→ 4. industrial
economic perspectives: 1. difficult to estimate→ 4. available industrial data
external use of energy: 1. difficult to decrease→ 4. no need
potential vol. of CO_2: 1. less than 10 Mton→ 4. more than 500 Mton
time of sequestration: 1. very short→ 4. long term
other impacts on env.: 1. significant→ 4. low (solvents or toxic. metals resources)

Fig. 3.7 Considerations about different CO_2 utilization: potential development, economic perspective, external use of energy, potential volume of CO_2, time of sequestration on other impact on environment (Ampelli et al. 2015)

3.3.3 CO_2 Storage Options

CO_2 can be stored into ground or ocean. In the first case, denominated as geological storage, CO_2 is injected into geological formations as depleted oil and gas reservoirs, deep saline aquifers, or for coal bed methane recovery and enhanced oil and gas recovery (they are both utilization and storage) at depths between 800 and 2000 m.

CO_2 is stored in "caprock" layers, that are impermeable trapping CO_2 (mudstones, clays, shales). According to the temperature and pressure of reservoir, CO_2 can be in different phase as gas, liquid or in supercritical conditions (31.1 °C and 73.8 bar) (Song and Zhang 2012). Deep saline aquifers have a storage capacity between 700 and 900 Gt CO_2 and can be offshore or onshore. Few information are present regarding coal bed methane recovery (Metz et al. 2005). On the other hand, a good knowledge is present about depleted oil and gas reservoirs. Figure 3.8 shows different geological storages.

The geological storage of CO_2 then uses the same technologies used for oil and gas industry and it is not economically feasible for unminable coal beds being at demonstration phase. Ocean can storage a huge amount of CO_2 at great depth, but it is not tested at large scale yet and it is at research phase (Li et al. 2013a, b). As shown in Fig. 3.9, two ocean storage options are present: the first consists on dissolving CO_2 into water at a deep below of 1000 m, by using fixed pipeline or a moving ship; the second consists on putting CO_2 via fixed pipeline or an offshore platform at sea floor, at a deep below of 3000 m, producing a lake due to the higher density than water.

CO_2 mineralization described in CO_2 utilization option can be also considered a CO_2 storage option.

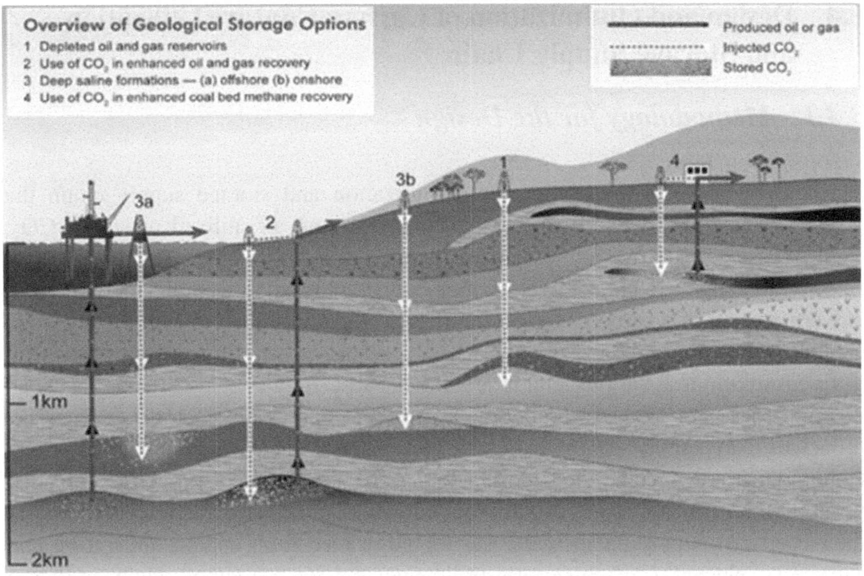

Fig. 3.8 Geological storage of CO_2 in depleted oil and gas reservoirs, for enhanced oil and gas recovery, in deep saline formations (onshore and offshore), for enhanced coal bed methane recovery (Metz et al. 2005)

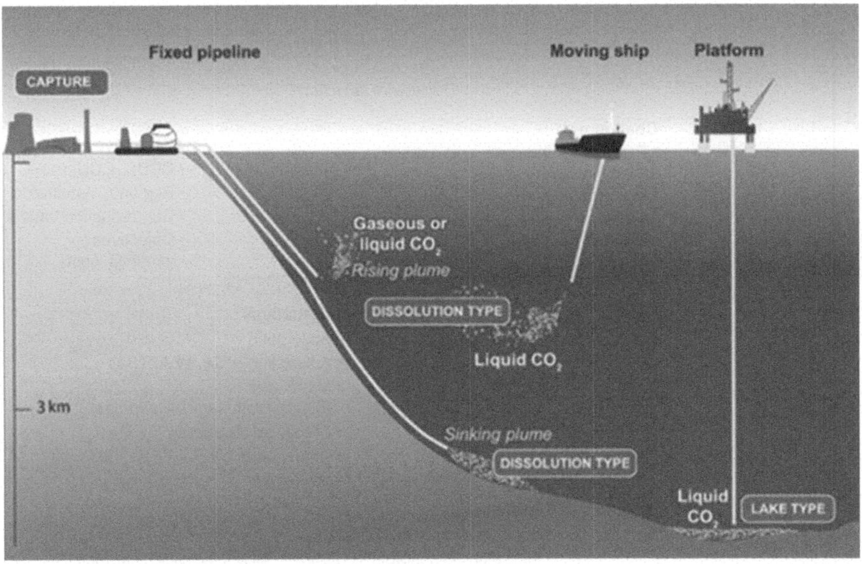

Fig. 3.9 Ocean storage of CO_2 with fixed pipeline (lake type storage) and moving ship (dissolution type storage) (Metz et al. 2005)

3.4 Design and Optimization of Carbon Capture Utilization and Storage Supply Chain

3.4.1 Methodology for the Design

Through the design of carbon capture utilization and storage supply chain the volume of captured CO_2 from flue gases, the volume of utilized or stored CO_2, right sources, capture technologies/materials, utilization and storage sites are selected. Then, different CO_2 emissions, geographic sites, capture technologies, etc. are present in a carbon supply chain. Therefore, as shown in Fig. 3.10, carbon capture utilization and storage supply chain design is a multi-scale problem with different solutions at various length- and time-scales (Biegler et al. 2014; Hasan et al. 2014, 2015; Lucia 2011). Overall costs are influenced by many factors, as materials, processes and supply chain level. At the material scale, materials for capture technologies are selected, considering their physical-chemical properties and disposability. At process scale, capture technologies are selected mainly according to their costs. At supply chain scale, the topology of carbon capture utilization and storage is selected. Carbon capture utilization and storage supply chain is then a complex and multi-scale system with performances depending on different factors.

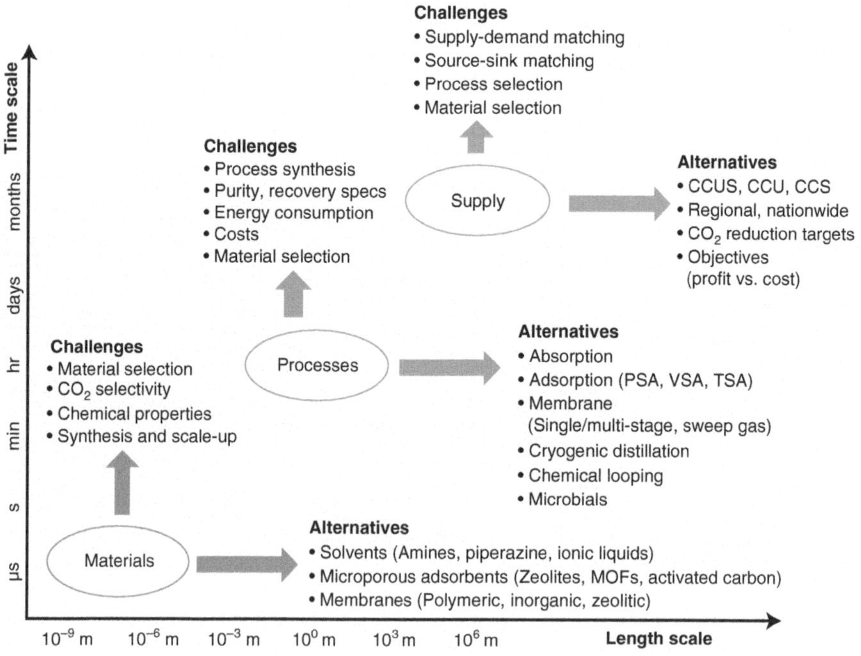

Fig. 3.10 Different scales of carbon capture utilization and storage supply chain design: material, processes and supply chain scales with challenges and alternatives (Hasan 2017)

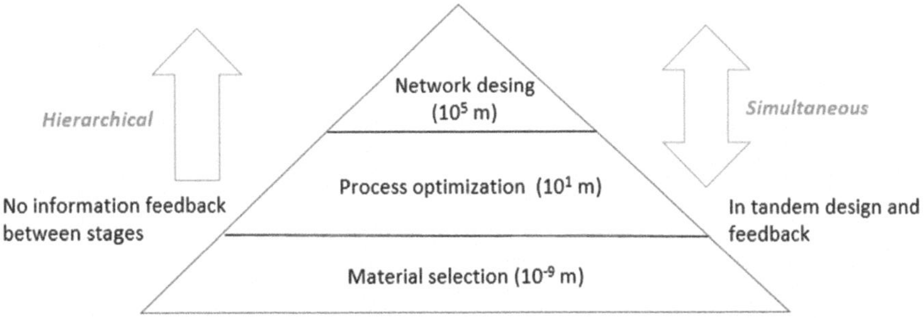

Fig. 3.11 Design pyramid for carbon capture utilization and storage supply chain: comparison between hierarchical (from bottom to top: material selection, process optimization and network design) and simultaneous approaches (tandem design of material selection, process optimization and network design) (Hasan 2017)

According to process systems engineering, two different approaches can be used in the design and optimization of multi-scale carbon capture utilization and storage supply chain at multiple length and time scales: hierarchical and simultaneous, as in Fig. 3.11. In hierarchical approach, information flows in only one direction of length scale: from bottom to the top of length scale. Also, decisions are made only at each scale (micro scale for material, meter scale for technology selection, kilometer scale for supply chain topology and optimization), then there is a suboptimal decision at each level that is passed from one scale to another in increasing sequence.

The following steps are identified by Hasan et al. (2015) as in Fig. 3.12: (i) materials screening, (ii) process optimization, (iii) process technology selection, (iv) materials screening and process optimization, (v) materials screening and process optimization and technology selection, and (vi) supply chain optimization. In this sequence, information of one step are used for the subsequent step that is in at first in micro-scale, then in meter-scale and at the end in kilometer scale.

In material screening step, the most appropriate capture material is selected according to its physical and chemical properties (First et al. 2011, 2013). In process optimization step, the best conditions are selected for the capture technology/material, chosen in the previous stage. These operating conditions are found in order to reduce capture and compression total costs also ensuring the specified purity and recovery. In process technology selection step, the capture technology is chosen by minimizing total costs. Generally, each capture technology can capture at least 90% of CO_2 that is compressed at 150 bar for utilization or storage. In material screening and process optimization step, the selection of capture materials and the optimization process that reduces total costs are developed in tandem, as in simultaneous way. In fact, carbon capture costs are influenced by CO_2 composition and flow rate and then, the choice found in the previous steps can be changed. It is important to optimize the system and to choose the best material simultaneously. In material screening, optimization and process technology selection step, different materials, processes and technologies are considered and optimized: the best technology is selected among other alternatives. The investment and operating costs of different capture

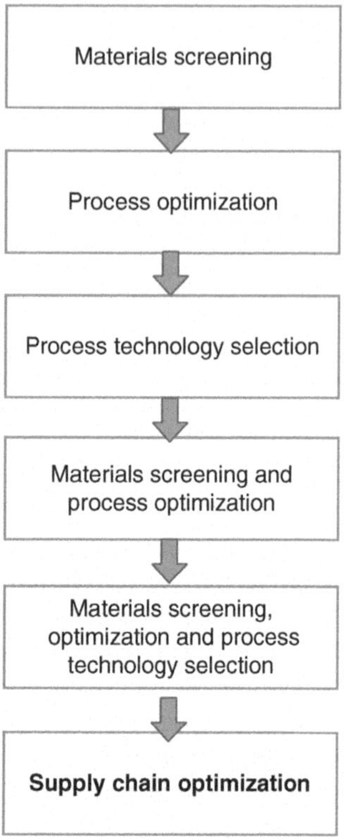

Fig. 3.12 A hierarchical framework for carbon capture utilization and storage supply chain design: at first materials screening, followed by process optimization, process technology selection, materials screening and process optimization, materials screening optimization and process technology, supply chain optimization (Hasan et al. 2015)

technologies/materials are expressed by the relation of Hasan et al. (2012a, b, 2014), as function of carbon composition, flue gas, and fixed parameters for each material/ technology. It is important to underline that no unique solution is present for the optimal solution of the process. Carbon flow rate and composition in the feed are important parameters in this context. Also, it is necessary to consider all configuration topology: all combination between CO_2 sources and capture technology must to be considered. The equation costs for carbon capture, expressed in following chapter, are an important link between process scale modeling and multiscale analysis of carbon capture utilization and storage supply chain network. In supply chain optimization step, the overall cost of supply chain is minimized. Total costs include CO_2 capture and compression costs, CO_2 transportation costs, CO_2 storage costs and CO_2 utilization costs (as the production costs of produced chemical compounds). In other cases, other economic parameters as net presentvalue, Pay-Back period, etc. can be considered for the optimization. Also, the objective function can describe an environmental parameter in order to maximize CO_2 reduction.

On the other hand, in simultaneous approach the design of supply chain is developed in tandem: the selection of material, process and supply chain is done in tandem, during the design of overall carbon capture utilization and storage supply chain. For this class of design, mixed integer linear programming (MILP) models are

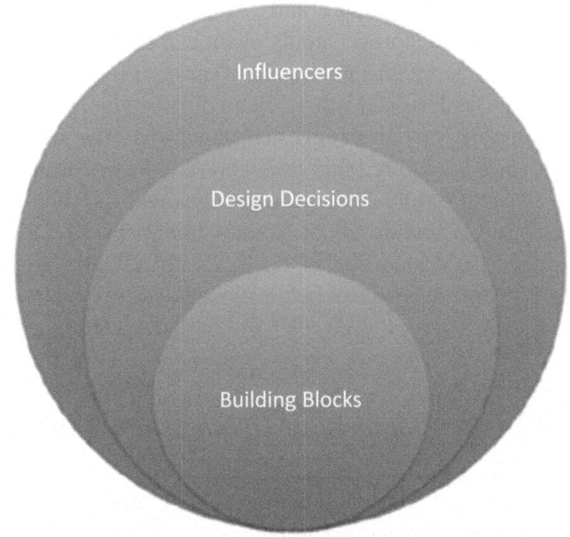

Fig. 3.13 Factors influencing a CCUS supply chain design: building blocks, design decisions and influencers (Melnyk et al. 2014)

carried out and they are able to select the best supply chain among multiple capture materials/technologies, CO_2 sources, utilization sites and storage sites. This methodology is developed because it is important to evaluate the overall supply chain for optimization problem; in fact, it is possible to have an optimum capture cost that is lower than that related to CO_2 capture alone. To this purpose, the connection between CO_2 source, CO_2 utilization and CO_2 storage sites is critical. Hasan et al. (2014) suggest at first the simultaneous design of carbon capture utilization and storage supply chain for Texas, where CO_2 is used to enhanced oil recovery.

Considering the two methods, however, the design of carbon capture utilization and storage supply chain, as other supply chain, should be developed considering critical interdependent factors such as influencers, design decisions, building blocks as reported in Fig. 3.13 (Melnyk et al. 2014). Political environment, business models and the supply chain life cycle can be considered as influencers. These factors influence the supply chain in a significative way. Social aspects and the physical/structural design of carbon capture utilization and storage supply chain are related to design decisions. Building blocks are related to aspects that are necessary to realize carbon capture utilization and storage supply chain.

All these considerations, suggest as the design of carbon capture utilization and storage supply chain is critical, strategic and complex. For this reason, Fine (1998) prefer the term "architecture" to design.

For the future, Melnyk et al. (2014) suggest some research area about the design of carbon capture utilization and storage supply chain: the dynamics design, the consideration of different outcomes, the stage of life cycle. It is important to consider not a static carbon capture utilization and storage supply chain but a dynamic one. Researchers should find and analyze factors that influence the dynamic process of supply chain. Melnyk et al. (2014), in their studies suggest that CCUS supply chain is increasingly dynamic and 'extreme'. For the second point, a carbon capture utilization and storage supply chain can be design putting different objective in the outcomes as increasing responsiveness, driving innovation or improving sustainability. The designed supply chain obtained by minimizing total costs should be not appropriate for these suggested objectives. The study of supply chain at different outcomes is required and few work are present in literature. For the last point, the presence of supply chain life is recognized by several work (Van Wassenhove 2006; Pettit and Beresford 2005; Kovács and Spens 2007), so it is possible analyzed and manage the transaction of one stage to next. As an important aspect a life cycle assessment analysis can be developed to verify that carbon dioxide loop is closed.

3.4.2 Development of Optimization Tool

To develop the design of carbon capture utilization and storage supply chain a tool for the optimization should be developed. Generally, A.I.M.M.S. (Advanced Interactive Multidimensional Modeling System) and G.A.M.S. (General Algebraic Modeling System) software are used. At first it is necessary to define a problem statement with:

- *assumptions*, that should be the following: capture plants are located at carbon dioxide sources, one to one coupling of source and capture node, carbon dioxide is transported via pipeline, the network is in stationary or dynamic conditions, the production of chemical compounds is constant over time, the technology used to produce the selected chemical compounds, etc.;
- *given information*, for example, about carbon dioxide source sites, carbon dioxide capture and compression technologies/materials, utilization and storage sites, the national demand of chemical produced compounds, distances between different considered sites, etc.;
- *conditions that must to be respected*, for example, the efficiency of carbon dioxide capture technologies, the maximum storage capacity of storage sites, the target of CO_2 reduction, etc.;
- *what should be decided* as CO_2 source, utilization and storage sites, then the topology, the volume of captured CO_2 that should be sent to utilization or storage, the amount of chemical compounds that are produced, the capture technology that should be used for each source, the best way to transport CO_2;
- *objective function,* for example total costs that should be minimized or CO_2 capture that should be maximize (other objective functions relating to economic aspect and environmental analysis can be also considered).

After that, it is necessary to define:

- *set* with a defined index for CO_2 sources, utilization and storage sites, capture technologies, etc.;

- *parameters* to define particular values required by the model, as the minimum target ofCO_2 reduction, maximum storage capacity, distance between different sites, CO_2 emissions for each source, flue gas flow rate and its CO_2 compositions, etc.;
- *variables*, binary (to select the combination between CO_2 source, capture technology/material and storage or utilization sites) or continues (to define the amount of CO_2 that is sent to utilization or storage, the amount of produced chemical compounds, etc.);
- *constraints* about the national demand of chemical produced compounds, the capacity of CO_2 storage sites, the selection of CO_2 capture technology/material, capture technology efficiency, the slitting of CO_2 sources, etc.
- *equations* for carbon dioxide capture and compression costs, transportation costs, storage costs, the production costs of different produced compounds in order to define the objective function.

Following these indications, a mathematical model of carbon capture utilization and storage supply chain should be developed and optimized due to the use of suggested software.

3.5 Cost Analysis

Generally, the total costs of carbon capture utilization and storage supply chain includes total carbon capture and compression costs, total transportation costs, total storage costs and the production costs of considered and produced chemical compounds. The evaluation of total carbon capture and compression costs is suggested by Hasan et al. (2014) according to the following relation (see Eq. 3.3):

$$CC = CDC + CIC + COC \tag{3.3}$$

where CC are total capture and compression costs, CDC are de-hydrogenation costs, CIC are investment costs while COC are operating costs. The de-hydrogenation is carried out by using tri-ethylene glycol absorption with a cost of 9.28 € per ton of CO_2 (including capital and investment costs) (Kalyanarengan Ravi et al. 2017). Investment and operating costs ($/year) are a function of CO_2 composition and flue gas flow are, respectively according to the following relations (see Eqs. 3.4 and 3.5):

$$CIC = \alpha_i + \left(\beta_i \cdot x_{co2}{}^{n_i} + \gamma_i\right) \cdot F^{m_i} \tag{3.4}$$

$$COC = \alpha_o + \left(\beta_o \cdot x_{co2,i}{}^{n_o} + \gamma_o\right) \cdot F^{m_o} \tag{3.5}$$

where α_i, α_o, β_i, β_o, γ_i, γ_o, n_i, n_o, m_i, m_o are fixed parameters depending on used technology and material, as in Table 3.3 (Zhang et al. 2018), x_{CO2} is CO_2 composition in flue gas and F is flue gas flow rate in mol/s. A trend of these costs as a

Table 3.3 Parameters of CO_2 capture and compression cost equations (investment and operation costs) (Zhang et al. 2018)

Process	Material	α	β	γ	n	m
Investment cost ($/year)						
Absorption	Monoethanolamine	7719	67,871	901.000	0.660	0.800
Absorption	Piperazine	0	59,956	226.932	0.566	0.800
Pressure swing adsorption	13X	220,462	26,720	895.262	0.508	0.804
Pressure swing adsorption	AHT	214,535	17,833	4607.297	0.744	0.813
Pressure swing adsorption	MVY	162,447	22,468	6408.791	1.000	0.797
Pressure swing adsorption	WEI	142,320	19,332	6076.357	0.610	0.779
Vacuum swing adsorption	13X	91,060	23,096	7688.408	0.470	0.763
Vacuum swing adsorption	AHT	113,969	24,939	2659.383	0.468	0.786
Vacuum swing adsorption	MVY	119,259	21,652	8101.014	1.000	0.795
Vacuum swing adorption	WEI	180,953	15,644	7751.257	0.874	0.802
Membrane	FSC PVAm	177,500	16,505	18912.000	0.880	0.770
Membrane	POE-1	568	19,151	29669.274	0.778	0.735
Membrane	POE-2	53,960	19,967	28462.417	0.656	0.744
Operating cost ($/year)						
Absorption	Monoethanolamine	0	24,088	0	1.000	1.000
Absorption	Piperazine	0	26,825	0	0.945	0.966
Pressure swing adsorption	13X	0	11,352	3115.833	1.000	0.974
Pressure swing adsorption	AHT	0	7040	983.893	0.626	1.000
Pressure swing adsorption	MVY	0	7265	1328.677	0.756	1.000
Pressure swing adsorption	WEI	0	6398	1257.721	0.554	0.991
Vacuum swing adsorption	13X	0	8167	1580.419	0.590	0.985
Vacuum swing adsorption	AHT	0	8545	1725.654	0.842	0.996
Vacuum swing adsorption	MVY	0	9117	1839.193	1.000	1.000
Vacuum swing adsorption	WEI	0	7378	1493.500	0.753	1.000
Membrane	FSC PVAm	0	11,619	0	0.210	1.000
Membrane	POE-1	0	12,798	0	0.134	0.980
Membrane	POE-2	0	13,556	0	0.135	0.984

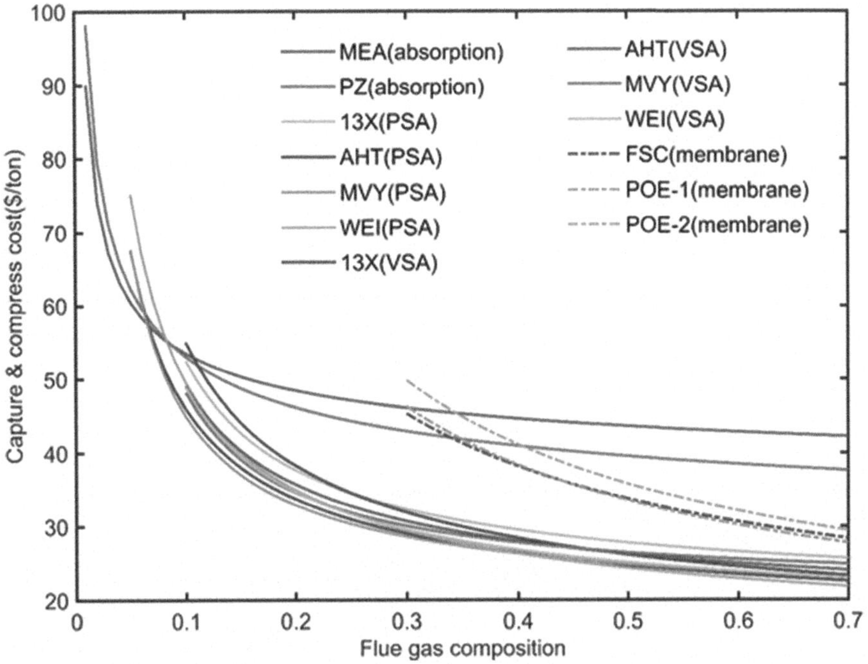

Fig. 3.14 CO_2 capture and compression costs for different capture technologies/materials with flue gas flow rate of 10 kmol/s (*MEA* monoethanolamine, *PZ* piperazine, *PSA* pressure swing adsorption, *VSA* vacuum swing adsorption; 13X, AHT, MVY, WEI are the materials for PSA and VSA system; FSC, POE-1, POE-2 are the materials for membrane system) (Zhang et al. 2018, reproduced with permission number 4563301123383)

function of carbon dioxide composition and flue gas flow rate for different technologies as absorption, pressure swing adsorption (PSA), membrane and vacuum swing adsorption (VSA) and materials as monoethanolamine (MEA), piperazine (PZ), AHT zeolite, MVY zeolite, 13X zeolite, FSC fixed site carried, POE1 polymer and PO2 polymer is reported in Fig. 3.14 (Zhang et al. 2018).

For ionic liquid absorption, different correlations ($/year) for investment and operative costs are respectively used, as the following equations (see Eqs. 3.6 and 3.7) (Nguyen et al. 2017):

$$CIC = (\alpha_i \cdot F + \beta_i) \cdot x_{co2} + \gamma_i \cdot F^{m_i} \tag{3.6}$$

$$COC = (\alpha_o \cdot F + \beta_o) \cdot x_{co2,i} + \gamma_o F^{mo} \tag{3.7}$$

where α_i, α_o, β_i, β_o, γ_i, γ_o, m_o, m_i are fixed parameters as in Table 3.4 (considering 1-butyl-3-methylimidazolium acetate [bmim][Ac] ionic liquid), x_{CO2} is carbon dioxide composition in flue gas, F is feed flow in mol/s.

Table 3.4 Parameters of CO_2 capture and compression cost equations with ionic liquid ([bmim] [Ac]) absorption (investment and operation costs) (Nguyen et al. 2017)

Process	Material	α	β	γ	m
Investment costs ($/year)					
Absorption	Ionic liquid	7.712	2,654,014	33546.87	0.67
Operating costs ($/year)					
Absorption	Ionic liquid	33172.59	897224.4	187421.2	0.65

Total costs for carbon dioxide transportation in pipeline is provided by Serpa et al. (2011) and Knoope et al. (2013): it is composed by investment and operating costs (€/year) as the following relation (See Eq. 3.8):

$$TC = TIC + TOC \tag{3.8}$$

where investment TIC (€/year) and operating costs TOC (€year) are respectively (see Eqs. 3.9 and 3.10):

$$TIC = (\alpha_t \cdot F_{co2} + \beta_t) \cdot F_T \cdot (D + F_c) \tag{3.9}$$

$$TOC = 4\% \cdot TIC \tag{3.10}$$

in the relations α_t is 0.019 and β_t is 0.533, D is distance between the considered sites and it is based on latitude and longitude (Kalyanarengan Ravi et al. 2017), F_T is a terrestrial factor (1.5 for mountain place, 2 for offshore, 1.4 for populated place, 1 for remote place) and generally an average value of 1.2 is considered (Broek et al. 2010), Fc is added to the distance to consider additional paths related to process (Dahowski et al. 2004), F_{CO2} is the amount of carbon dioxide that is transported. Among other way of CO_2 transportation, pipeline is the favorite choice, being the most mature solution. Also, pipeline ensures the transportation of large volume of CO_2 at low costs. (Kalyanarengan Ravi et al. 2017).

Total carbon dioxide storage costs (CS) comprise investment costs (SIC) and operating costs (SOC), as in the following relation (Ochoa Bique et al. 2018) (see Eq. 3.11):

$$CS = SIC + SOC \tag{3.11}$$

where investment costs (€/year) are calculated as (Hendriks 1994) (see Eq. 3.12):

$$SIC = (m \cdot d_{well} + b) \cdot N_{well}^{build} \tag{3.12}$$

where m and b are parameters respectively corresponding to $1.53 \cdot 10^6$ and $1.23 \cdot 10^6$ (Ochoa Bique et al. 2018), d_{well} is the depth of well storage and N_{well}^{build} is the number of wells built per year (Hasan et al. 2014) (see Eq. 3.13):

$$N_{well}^{build} = \frac{F_{s,co2}}{IC} \tag{3.13}$$

that is a function of stored carbon dioxide $F_{s,CO2}$ and injection capacity per well IC. Operating costs (€/year) are 4% of investment costs (see Eq. 3.14):

$$SOC = 4\% \cdot SIC \tag{3.14}$$

The production costs of each chemical compound can be found in literature.

3.6 Literature Work About Carbon Capture Utilization and Storage Supply Chain

Carbon capture utilization and storage supply chain can be considered as a long-term CO_2 containment and despite its important role for the environment not many work are reported in literature about it.

O'Brien et al. (2017) suggest that carbon capture utilization and storage supply chains have a decisive role for the economic development. To this purpose can be useful to have a methodology to evaluate large scale capture options, to consider different CO_2 utilization options that are relevant for local economy and to establish a public-private partnership incentivizing carbon supply chains. Different models for the State of Illinois are presented. The work also suggests the actions required for the workforce development throughout the carbon supply chain.

Suggestions, recommendations and policies for the development of carbon capture utilization and storage supply chains in China as an important and strategic instrument to reduce CO_2 emissions and to ensure energy security can be found in the work of Zhang et al. (2013a, b).

Considerations about carbon capture utilization and storage supply chain are reported by Floudas and Nye (2015) reducing up to 50% the total stationary CO_2 emissions. The authors underline as total costs vary with CO_2 capture materials/ technologies, selected CO_2 sources, utilization and storage sites, CO_2 transportation, the amount of stored CO_2. Regarding this aspect, to understand the economy of CCUS supply chain, each component should be evaluated separately or integrated in all systems. Vikara et al. (2017) present tools, models, resources to this purpose.

Han and Lee (2012) propose a multi-period stochastic model for a carbon capture utilization and storage system in Korea to manage CO_2 emissions from 2011 to 2030. The model is a MILP and the uncertainty of price, costs and CO_2 emissions are considered. Results show that considering only the uncertainty of CO2 emissions good results are obtained as those obtained considering all uncertainties. The work is subsequently improved in Han and Lee (2013) considering different aspects as techno-economic-, environmental- and risk.

The selection of right technology and respective CO_2 source is critical to develop a profitable carbon capture utilization and storage supply chain, as well as the selection of technologies and materials (Hasan et al. 2013a). Mohd Rudin et al. (2017) develop a new strategy suggesting the reduction of emissions from early stages. A methodology of carbon reducing, capture, utilization and storage known as CARCUS is proposed. Considering the reduction of CO_2 in the early stage, capture costs can be reduced because emissions are reduced earlier and smaller equipments for capture can be utilized. A mixed integer linear programming model is obtained and applied for Malaysia region. The best technologies that should be used are suggested.

CO_2 enhanced oil recovery is considered as a good option to use CO_2, then several carbon capture utilization and storage supply chains with CO_2 enhanced oil recovery are analyzed in literature. This solution can rise oil recovery by 15% of original oil in place (OOIP) (National Energy Technology Laboratory 2010a). Kuuskraa et al. (2013), analyzing CO_2-EOR technology in the residual oil field in Texas, find that revenues obtained by the selling of captured CO_2 can hurry the application of this kind of carbon capture utilization and storage supply chain at a wide scale. CO_2 is tight stored: it is evident that CO_2 enhanced oil recovery will have an important role in worldwide carbon management strategy and in the related supply chain. A benefit-risk analysis of these systems for two stakeholders (power plant owner and oil field owner) is carried out by Zhu and Liu (2015): oil field has the main position in CO_2 utilization while contract designs have to get better the position of power plants. A model for carbon capture utilization and storage supply chains should be developed in order to gain insights for planning CO_2 utilization network.

Hasan et al. (2013b) optimize with a mixed integer programming (MIP) model a carbon capture utilization and storage supply chain for national and regional level in the United States. Total expenditure is minimized with a fixed CO_2 reduction target. The model is able to select source plants, capture systems (absorption, membrane, pressure swing adsorption and vacuum swing adsorption), capture materials (2 solvents, 4 zeolites, and 3 membranes), locations of utilization (oil and gas reservoirs) and storage sites. Results show that decreasing 50% of CO_2 emissions the cost is lower than 30$ per ton of CO_2, an optimal value. It is underlined also as the choice of capture technology/material is crucial to be considered inside large-scale carbon capture utilization and storage design.

Hasan et al. (2014) suggest a carbon capture utilization and storage supply chain model that can be applied at national, regional and statewide level in the United State where CO_2 is used to enhanced oil recovery. It is the first model that considers the costs of CO_2 capture, compression, transportation, utilization and sequestration and at the same time choices CO_2 sources, capture materials, capture technologies, utilization and sequestration sites minimizing total costs. The authors, also, present several correlations to evaluate the capture costs of different capture systems, as a function of gas composition and flow rate. The authors find that the system is able to decrease 50–80% of CO_2 emissions with a total costs between $58.1–106.6 billion, generating $3.4–3.6 billion of revenues. The specific cost of carbon capture utilization and storage supply chain is $35.63–43.44 per ton of captured CO_2.

Hasan et al. (2015) develop a strategy for the optimal design of supply chain minimizing total costs and reducing CO_2 emissions in the United Stated. In utilization section, CO_2 is used to recover oil. The overall framework includes the following steps: materials screening, process optimization, process technology selection, materials screening and process optimization, materials screening and process optimization and technology selection, and supply chain optimization. Information are passed by one previous step to another following step. The system is described by a mixed inter programming model. 444 sources for CO_2 capture, 76 oil and gas reservoirs for CO_2 utilization, 151 saline formations and 6 unmineable coalbeds for CO_2 sequestration are present in the analyzed system. For the capture, 13 technology-material pairs (monoethanolamine and piperazine for absorption; FSCPVAm, POE1 and POE2 for membrane; 13X, AHT, MVY and WEI for pressure swing adsorption; 13X, AHT, MVY and WEI for vacuum swing adsorption) are considered. Authors suggest that the optimization of supply chain is a combinatorial problem where the optimal solutions are increasing with storage, utilization and source sites. The system is optimized minimizing total costs. 50% of CO_2 emission are reduced at $35.63 per ton of captured CO_2. Results show that the cost of supply chain increases with minimum CO_2 reduction. In fact, reducing 60% and 80% CO_2, the costs of supply chain are equal to 36.93$ and 43.44$ respectively. In this analysis, the costs of dehydration, capture and compression have a higher influence on total costs, due to the high value these costs represent an impediment for the development of these technologies.

Tapia et al. (2016a) present a mixed integer linear programming model with discrete time optimization approach for carbon capture utilization and storage supply chain with CO_2 enhanced oil recovery as utilization. In the model, respect to other literature work, CO_2 allocation and scheduling issues are considered for EOR operation. Only one source is considered while more different reservoirs are present. In addition, two case studies are evaluated: one case study considers a fixed amount of CO_2, while in another case study CO_2 flow rate is varying during the considered period. The authors suggest to consider in future work several uncertainties as: oil price, reservoirs oil capacity and oil yield through Monte Carlo simulation or sensitivity analysis.

More sources are considered in Tapia et al. (2016b). Also, in this work a carbon capture utilization and storage supply chain with CO_2 enhanced oil recovery is considered, but compared to other literature work, three important considerations for the development of these kind of systems are underlined: scheduling of carbon capture utilization and storage operation, allocation of CO_2 supply for enhanced oil recovery operation and matching CO_2 source and geological sinks. Two mixed integer linear programming models are carried out: one for the design of an enhanced oil recovery system by using strip packing analogy (Castro and Grossmann 2012) and another for source–sink matching in carbon capture storage systems. These two models are presented and described using a case study. Results for the first model show a direct proportionality between the amount of CO_2 in flue gas and profit. In the second model, results show that there is a no linear proportionality between the amount of captured and stored CO_2 and injectivity limits, but the a direct proportionality is present between the amount of capture and stored CO_2 and the total available capacities.

Middleton et al. (2015) develop an interesting carbon capture utilization and storage supply chain: ethylene plants are CO_2 sources and CO_2 is used to recover oil in the US Gulf Coast region and in near regions. This integration can increase the production cost in a lower measure than the case of CO_2 captured by fossil power plant: the increase of ethylene price ranges between 1% and 15%, while the increase of electricity price ranges between 50% and 100% with CO_2 captured by fossil fuel power plant. The network can reduce 50 $MtCO_2$/year emissions producing 200 million bbl/year of oil. Then system is economically feasible then can increase its social acceptance as a climate change mitigation technology.

Other mathematical models for carbon capture utilization and storage supply chain with enhanced oil recovery as CO_2 utilization are developed by Rahmawati et al. (2015) improving the work of Hassiba et al. (2016) by considering heat integration in the system.

Klokk et al. (2010) suggest a mathematical model for carbon capture utilization and storage supply chain with CO_2 enhanced oil recovery for Norwegian region with 14 oil fields, two aquifers and five CO_2 sources. Considering a fixed amount of CO_2 emissions, the system is optimized maximizing net present value, then selecting the best oil fields and geological storages. A parametric study is developed considering the most important factors. Results also suggest that the use of CO_2 for the analyzed scope can determine a profitable system in economic sense, even if expensive infrastructure is used.

Romanenko (2014) develop a system dynamic model for a carbon capture utilization and storage supply chain with enhanced oil recovery as CO_2 utilization, analyzing the influence of different policy designs. In this context, the Carbon Tax Credit policy determines several advantages on the system allowing its self-sustaining growth. This policy is defined as reinforcing mechanism.

Sun and Chen (2017) analyze different carbon capture utilization and storage supply chain with CO_2 enhanced water recovery (CO2-EWR), CO_2 enhanced oil recovery (CO_2-EOR) and CO_2 enhanced coalbed methane recovery (CO_2-ECBM) in China. Results show that in a short-period CO_2 enhanced oil recovery is the most favorable carbon capture utilization and storage technology while in a long term CO_2 enhanced water recovery is the best solution because saline aquifer has the largest storage capacity. The authors underline as for this kind of technology financial and political supports are required for their application in order to provide a great contribution to low carbon economy not only for China but for all the word.

A more detailed economic analysis for CO_2 enhanced oil recovery system is developed by Kwak and Kim (2017): net present value is optimized setting some constrain on dynamic CO_2 sources. A sensitivity analysis is developed in order to analyze the influence of oil and CO_2 price on total costs. Also, situations with a higher and lower amount of CO_2 compared to base case are evaluated. It is found that reducing CO_2 provision of 20%, overall oil production is decreased of 33.7%, otherwise increasing the CO_2 supply chain rate of 20%, overall oil production is increased of 5.6%.

Wu et al. (2015) provide an optimized model for carbon capture utilization and storage supply chain with enhanced oil recovery as CO_2 utilization under uncertainty about technical, economic and political aspects applied for China case. Results suggest that carbon price can have a political influence for the development of these technologies. Without a high cost of CO_2, it is difficult to realize this system in a short time.

Other utilization cases can be considered inside carbon capture utilization and storage supply chain. Yue and You (2015) develop a multi-scale multi-period mixed-integer nonlinear programming (MINLP) model for carbon capture utilization and storage supply chain in Texas where captured CO_2 is used by algae converting CO_2 into lipids, that are converted into renewable biofuels. Through Life Cycle Optimization (LCO) methodology the system is optimized minimizing CO_2 reduction costs and maximizing the amount of CO_2 avoided. A good reduction of CO_2 is achieved and equal to 80% producing 187 Mgal/year of renewable diesel. However, only the utilization of captured CO_2 is not able to obtain 80% of CO_2 reduction.

Ochoa Bique et al. (2018) study a carbon capture utilization and storage supply chain in Germany considering methanol production in the utilization site. CO_2 is captured from power plants while hydrogen for the reaction producing methanol is obtained by water electrolysis. The integration of CO_2 and H_2 supply chain is then evident. Results show that the system can be profitable if renewable hydrogen is distributed without paying.

Leonzio et al. (2019) suggest a methodology for the optimal design of carbon capture utilization and storage supply chain in Germany: the captured CO_2 is used to produce methanol via methane dry reforming. The supply chain is optimized minimizing total costs, using Advanced Interactive Multidimensional Modeling System (AIMMS) software. Results show that the system is economic feasible only considering economic incentives. Also, Germany will have, in the next future, an important role in world methanol market.

Other literature work pay attention on other aspects of carbon capture utilization and storage supply chain. Sun and Chen (2017) present a source-sink matching multi-stage programming model to design CO_2 pipeline layout using General Algebraic Modeling System (GAMS) software. It is, in particular, a mixed integer programming model minimizing net present value. Carbon capture utilization and storage supply chain is located in China and in particular in Jing-Jin-Ji region (Bohai Sea area). Results show that to transport 1620 Mt. CO_2 during 2020–2050 period, 2200 km of pipeline are required with and investment cost of $1.6 billion. The developed model is able to determine CO_2 transportation structure (number and length of pipeline), location and amount of captured CO_2. The authors suggest that in future work it is necessary to integrate an energy system optimization model to study a more realistic system.

From these literature work it is evident that mathematical programming is used in the planning of large-scale carbon capture utilization and storage supply chain, providing a flexible and high resolution. This is very important, because planning methods and tools are related to political decisions for carbon capture utilization and storage (Bryngelsson and Hansson 2009; Tapia et al. 2018). Mathematical techniques are classified according to the method used for the design, in energy models, pipeline infrastructure design and source-sink matching (SSM). First models are related to the energy balance of carbon capture utilization and storage systems, considering power losses during capture phase (Tapia et al. 2018). Pipeline infrastructure design model is related to the economics of construction and management for CO_2 distribution networks. This aspect is very important, because the transportation cost of captured CO_2 contributes with 10% to total cost in carbon capture utilization and storage supply chain (Rubin et al. 2015). A good design of carbon capture utilization and storage pipeline network avoids the reconstruction or expansion after the realization that require a great investment. As also shown previously, single-stage and multi-stage models are carried out for carbon capture utilization and storage pipeline networks in China (Sun and Chen 2015, 2017): pipeline network is optimized minimizing CO_2 transportation costs. Source-sink matching model is related to find CO_2 sources that should be matched with different utilization or storage options, without considering pipeline networks. In this model, source and sink are matched in terms of quality and continuity by optimizing (Tapia et al. 2016a, 2016b).

However, it is necessary to keep in mind and put more attention on the uncertainties of these models deriving by decades-long time horizons, as climatic and economic conditions, geophysical properties of storage reservoirs and environmental policies. Uncertainties are due to the dynamic nature of supply chain as also reported by Ozkır and Baslıgil (2013). For this reason, in the future, more mathematical models considering uncertainty should be developed. A similar work is also developed already and in a more general way for green supply chain network by Rahmani and Mahoodian (2017) considering the uncertainty risk of some parameters (demand and facilities). These indicated risks are suggested by Hatefi and Jolai (2014). Then, these considerations can be more extended to carbon capture utilization and storage supply chains in the next researches in order to analyze a more realistic problem.

3.7 Conclusions

In this analysis an overview about carbon capture utilization and storage supply chain is proposed. In these systems CO_2 is captured, transported and sent to utilization and/or storage sites and it is the only system that can allow to achieve the objectives of environmental policy established in the Paris Climate Change Agreement. Several plants are present in the world, actually and others should be realized. For the important role of them, mathematical modellings with respective equations should be developed for optimization and design. Generally, a hierarchical and simultaneous methods can be used. A strategy for the modeling and optimization of carbon capture utilization and storage supply chain is also suggested minimizing total costs. The work presents, in addition, an overview about CO_2 capture technology, CO_2 storage and CO_2 utilization. CO_2 can be captured by chemical/physical absorption, adsorption, membrane, chemical looping combustion, cryogenic and hybrid technologies. Advantages and disadvantages are reported. CO_2 can be stored in geological storage or in ocean. CO_2 can be used directly, as injection fluid for oil or gas recovery, to produce different chemical compounds and fuels or for mineral carbonation. In literature many work are present about the modeling of carbon capture utilization and storage supply chains. However more studies should be done considering dynamic systems with uncertainties.

Acknowledgements The author would like to thank the University of L'Aquila for funding this work.

References

Abatzoglou N, Boivin S (2009) A review of biogas purification processes. Biofuels Bioprod Biorefin 3:42–71. https://doi.org/10.1002/bbb.117

Ampelli C, Perathoner S, Centi G (2015) CO2 utilization: an enabling element to move to a resource- and energy-efficient chemical and fuel production. Phil Trans R Soc A 373:20140177. https://doi.org/10.1098/rsta.2014.0177

AxeL M, Xiaoshan S (1997) Research and development issues in CO2 capture. Energy Convers Manag 38:37–42. https://doi.org/10.1016/S0196-8904(96)00242-7

Bamdad H, Hawboldt K, MacQuarrie S (2018) A review on common adsorbents for acid gases removal: focus on biochar. Renew Sust Energ Rev 81:1705–1720. https://doi.org/10.1016/j.rser.2017.05.261

Bhown AS (2014) Status and analysis of next generation post-combustion CO2 capture technologies. Energy Procedia 63:542–549. https://doi.org/10.1016/j.egypro.2014.11.059

Biegler LT, Lang YD, Lin WJ (2014) Multi-scale optimization for process systems engineering. Comput Chem Eng 60(10):17–30. https://doi.org/10.1016/j.compchemeng.2013.07.009

Broek MVD, Brederode E, Ramírez A, Kramers L, Kuip MVD, Wildenborg T, Turkenburg W, Faaij A (2010) Environmental modelling & software designing a cost-effective CO2 storage infrastructure using a GIS based linear optimization energy model. Environ Model Softw 25 (12):1754–1768. https://doi.org/10.1016/j.envsoft.2010.06.015

Brunetti A, Scura F, Barbieri G, Drioli E (2010) Membrane technologies for CO2 separation. J Membr Sci 359:115–125. https://doi.org/10.1016/j.memsci.2009.11.040

Bryngelsson M, Hansson A (2009) Energy policy on shaky ground? A study of CCS-scenarios. Energy Procedia 1:4673–4680. https://doi.org/10.1016/j.egypro.2009.02.290

Castro PM, Grossmann IE (2012) From time representation inscheduling to the solution of strip packing problems. ComputChem Eng 44:45–57. https://doi.org/10.1016/j.compchemeng.2012.05.002

Cuellar-Franca RM, Azapagic A (2015) Carbon capture, storage and utilisation technologies: a critical analysis and comparison of their life cycle environmental impacts. J CO2 Util 9:82–102. https://doi.org/10.1016/j.jcou.2014.12.001

Dahowski R, Dooley J, Davidson C, Bachu S, Gupta N (2004) A CO2 storage supply curve for North America. In: Proceedings of the 7th international conference on greenhouse gas control technologies

Dai Z, Ansaloni L, Deng L (2016) Recent advances in multi-layer composite polymeric membranes for CO2 separation: a review. Green. Energy Environ 1:102–128. https://doi.org/10.1016/j.gee.2016.08.001

Evangelos PF, Fotios KK, Sergios KP, Andreas AS, Athanasios CM (2017) A review of the latest development of polyimide based membranes for CO2 separations. React Funct Polym 120:104–130. https://doi.org/10.1016/j.reactfunctpolym.2017.09.002

Figueroa JD, Fout T, Plasynski S, McIlvried H, Srivastava RD (2008) Advances in CO2 capture technology—the U.S. Department of Energy's Carbon Sequestration Program. Int J Greenhouse Gas Control 2:9–20. https://doi.org/10.1016/S1750-5836(07)00094-1

Fine CH (1998) Clockspeed: winning industry control in the age of temporary advantage. Perseus Books, Reading

First EL, Gounaris CE, Wei J, Floudas CA (2011) Computational characterization of zeolite porous networks: an automated approach. Phys Chem Chem Phys 13:17339–17358. https://doi.org/10.1039/C1CP21731C

First EL, Gounaris CE, Floudas CA (2013) Predictive framework for shape-selective separations in three-dimensional zeolites and metal–organic frameworks. Langmuir 29(18):5599–5608. https://doi.org/10.1021/la400547a

Floudas CA, Nye E, (2015) Carbon capture, utilization, and sequestration: a multi-scale grand challenge. http://media.journals.elsevier.com/content/files/floudasdanckwertsnice-13105423.pdf

Gao W, Zhou T, Gao Y, Louis B, O'Hare D, Wang Q (2017) Molten salts-modified MgObased adsorbents for intermediate-temperature CO2 capture: a review. J Energy Chem 26:830–838. https://doi.org/10.1016/j.jechem.2017.06.005

Gupta M, Coyle I, Thambimuthu K (2003) Strawman document for CO2 capture and storage technology roadmap. Canmet Energy Technology Centre, Natural Resources, Canada. http://www.graz-cycle.tugraz.at/pdfs/co2_capture_strawman_feb2004.pdf

Han JH, Lee IB (2011) Development of a scalable and comprehensive infrastructure model for carbon dioxide utilization and disposal. Ind Eng Chem Res 50:6297–6631. https://doi.org/10.1021/ie200344t

Han JH, Lee IB (2012) Multiperiod stochastic optimization model for carbon capture and storage infrastructure under uncertainty in CO2 emissions product prices, and operating costs. Ind Eng Chem Res 51:11445–11457. https://doi.org/10.1021/ie3004754

Han JH, Lee IB (2013) A comprehensive infrastructure assessment model for carbon capture and storage responding to climate change under uncertainty. Ind Eng Chem Res 52:3805–3815. https://doi.org/10.1021/ie301451e

Hasan MMF (2017) Multi-scale process systems engineering for carbon capture, utilization, and storage: a review, process systems and materials for CO2 capture: modelling, design, control and integration, 1st edn. Edited by Athanasios I. Papadopoulos and Panos Seferlis. Wiley.

Hasan MMF, Baliban RC, Elia JA, Floudas CA (2012a) Modeling, simulation, and optimization of postcombustion CO2 capture for variable feed concentration and flow rate. 1. Chemical absorption and membrane processes. Ind Eng Chem Res 51(48):15642–15664. https://doi.org/10.1021/ie301571d

Hasan MMF, Baliban RC, Elia JA, Floudas CA (2012b) Modeling, simulation, and optimization of postcombustion CO2 capture for variable feed concentration and flow rate.2. Pressure swing adsorption and vacuum swing adsorption processes. Ind Eng Chem Res 51(48):15665–15682. https://doi.org/10.1021/ie301572n

Hasan MMF, First EL, Floudas CA (2013a) Cost-effective CO2 capture based on in silico screening of zeolites and process optimization. Phys Chem Chem Phys 15(40):17601–17618. https://doi.org/10.1039/C3CP53627K

Hasan MMF, Boukouvala F, Floudas CA (2013b) Optimization of CO_2 capture, utilization and sequestration (CCUS) supply chain networks, AIChE Annual Meeting in San Fransisco November 08, 2013

Hasan MMF, Boukouvala F, First EL, Floudas CA (2014) Nationwide, regional and statewide CO2 capture, utilization and sequestration supply chain network optimization. Ind Eng Chem Res 53 (18):7489–7506. https://doi.org/10.1021/ie402931c

Hasan MMF, First EL, Boukouvala F, Floudas CA (2015) A multi-scale framework for CO2capture, utilization, and sequestration: CCUS and CCU. Comput Chem Eng 81:2–21. https://doi.org/10.1016/j.compchemeng.2015.04.034
Hassiba RJ, Al-Mohannadi DM, Linke P (2016) Carbon dioxide and heat integration of industrial parks. J Clean Prod 155(1):47–56. https://doi.org/10.1016/j.jclepro.2016.09.094
Hatefi SM, Jolai F (2014) Robust and reliable forward-reverse logistics network design under demand uncertainty and facility disruptions. Appl Math Model 38(9):2630–2647. https://doi.org/10.1016/j.apm.2013.11.002
Hedin N, Andersson L, Bergstr ML, Yan J (2013) Adsorbents for the post-combustion capture of CO2 using rapid temperature swing or vacuum swing adsorption. Appl Energy 104:418–433. https://doi.org/10.1016/j.apenergy.2012.11.034
Hendriks CA (1994) Carbon dioxide removal from coal-fired power plant. In Department of Science. Technology and Society. Utrecht University, Utrecht
Hussain A, Hägg MB (2010) A feasibility study of CO2 capture from flue gas by a facilitated transport membrane. J Membr Sci 359:140–148. https://doi.org/10.1016/j.memsci.2009.11.035
Kalyanarengan Ravi N, Van Sint Annaland M, Fransoo JC, Grievink J, Zondervan E (2017) Development and implementation of supply chain optimization framework for CO2 capture and storage in the Netherlands. Comp Chem Eng 102:40–51. https://doi.org/10.1016/j.compchemeng.2016.08.011
Klokk O, Schreiner PF, Pages-Bernaus A, Tomasgard A (2010) Optimizing a CO_2 value chain for the Norwegian continental shelf. Energy Policy 38:6604–6614. https://doi.org/10.1016/j.enpol.2010.06.031
Knoope MMJ, Ramrez A, Faaij APC (2013) A state-of-the-art review oftechno-economic models predicting the costs of CO2pipeline transport. Int J Greenhouse Gas Control 16:241–270. https://doi.org/10.1016/j.ijggc.2013.01.005
Kovács G, Spens KM (2007) Humanitarian logistics in disaster relief operations. Int J Phys Distrib Logist Manag 37(2):99–114. https://doi.org/10.1108/09600030710734820
Kuuskraa VA, Godec ML, Dipietro P (2013) CO2 utilization from "next generation" CO2 enhanced oil recovery technology. Energy Procedia 37:6854–6866. https://doi.org/10.1016/j.egypro.2013.06.618
Kwak D, Kim JK (2017) Techno-economic evaluation of CO2enhanced oil recovery (EOR) with the optimization of CO2 supply. Int J Greenhouse Gas Control 58:169–184. https://doi.org/10.1016/j.ijggc.2017.01.002
Leonzio G (2018) State of art and perspectives about the production of methanol, dimethyl ether and syngas by carbon dioxide hydrogenation. J CO2 Util 27:326–354. https://doi.org/10.1016/j.jcou.2018.08.005
Leonzio G, Foscolo PU, Zondervan E (2019) An outlook towards 2030: optimization and design of a CCUS supply chain in Germany. Comp Chem Eng 125:499–513. https://doi.org/10.1016/j.compchemeng.2019.04.001
Leung DYC, Caramanna G, Maroto-Valer MM (2014) An overview of current status of carbon dioxide capture and storage technologies. Renew Sust Energ Rev 39:426–443. https://doi.org/10.1016/j.rser.2014.07.093
Li B, Duan Y, Luebke D, Morreale B (2013a) Advances in CO2 capture technology: a patent review. Appl Energy 102:1439–1447. https://doi.org/10.1016/j.apenergy.2012.09.009
Li L, Zhao N, Wei W, Sun Y (2013b) A review of research progress on CO2 capture, storage, and utilization in Chinese Academy of Sciences. Fuel 108:112–130. https://doi.org/10.1016/j.fuel.2011.08.022
Liu HJ, Were P, Li Q, Gou Y, Hou Z (2017) Worldwide status of CCUS technologies and their development and challenges in China. Geofluids 2017:1–25. https://doi.org/10.1155/2017/6126505
Lucia A (2011) Multi-scale methods and complex processes: a survey and look ahead. Comput Chem Eng 34:1467–1475. https://doi.org/10.1016/j.compchemeng.2009.10.004

Luo X, Wang C (2017) The development of carbon capture by functionalized ionic liquids. Curr Opin Green Sustain Chem 3:33–38. https://doi.org/10.1016/j.cogsc.2016.10.005

Ma Y, Gao J, Wang Y, Hu J, Cui P (2018) Ionic liquid-based CO2 capture in power plants for low carbon emissions. Int J Greenhouse Gas Control 75:134–139. https://doi.org/10.1016/j.ijggc.2018.05.025

Mamun S, Dindore VY, Svendsen HF (2007) Kinetics of the reaction of carbon dioxide with aqueous solutions of 2-((2-aminoethyl) amino) ethanol. Ind Eng Chem Res 46:385–394. https://doi.org/10.1021/ie060383v

Melnyk SA, Narasimhan R, DeCampos HA (2014) Supply chain design: issues, challenges, frameworks and solutions. Int J Prod Res 52:1887–1896. https://doi.org/10.1080/00207543.2013.787175

Metz B, Davidson O, de Coninck H, Loos M, Meyer L (2005) IPCC special report on carbon dioxide capture and storage, Intergovernmental Panel on Climate Change. Available at: http://www.ipcc.ch/pdf/special-reports/srccs/srccs_wholereport.pdf

Middleton RS, Levine J, Bielicki J, Carey JW, Viswanathan HS, Stauffer PH (2015) Jumpstarting commercial-scale CO2 capture and storage with ethylene production and enhanced oil recovery in the U.S. Gulf. Greenhouse Gases Sci Technol 5(3):241–253

Milani D, Khalilpour R, Zahedi G, Abbas A (2015) A model-based analysis of CO2 utilization in methanol synthesis plant. J CO2 Util 10:12–22. https://doi.org/10.1016/j.jcou.2015.02.003

Mohd Rudin SNF, Muis ZA, Hashim H, Ho WS (2017) Overview of carbon reduction, capture, utilization and storage: development of new framework. Chem Eng Trans 56:649–654. https://doi.org/10.3303/CET1756109

Mondal MK, Balsora HK, Varshney P (2012) Progress and trends in CO2 capture/separation technologies: a review. Energy 46:431–441. https://doi.org/10.1016/j.energy.2012.08.006

Nandi S, De Luna P, Daff TD, Rother J, Liu M, Buchanan W et al (2015) A single-ligand ultra-microporous MOF for pre-combustion CO2 capture and hydrogen purification. Sci Adv 1(11): e150042. https://doi.org/10.1126/sciadv.1500421

National Energy Technology Laboratory (2010a) DOE/NETL carbon dioxide capture and Storage RD&D Roadmap. US Department of Energy, Washington, DC

National Energy Technology Laboratory (2010b) NETL 2010 carbon sequestration atlas of the United States and Canada. 3rd ed. Atlas III. Available at www.netl.doe.gov/technologies/carbonseq/refshelf/atlas

Nguyen TBH, Reisemann SG, Zondervan E (2017) Development of a conceptual process for CO_2 capture from flue gases using ionic liquid. Proceedings of the 27th European Symposium on Computer Aided Process Engineering – ESCAPE 27 October 1st – 5th. 2017. Barcelona. Spain https://doi.org/10.1016/B978-0-444-63965-3.50439-6

Nsakala N, Liljedahl GN (2003) Greenhouse gas emissions control by oxygen firing in circulating fluidized bed boilers: phase 1 – a preliminary systems evaluation. Final report, U.S. Department of Energy, DE-FC26-01NT41146

O'Brien K, Patel V, Lu Y, Krishnamurthy KR (2017) CCUS as a regional economic development tool: planning and design considerations, carbon management technology conference (CMTC 2017)

Ochoa Bique A, Nguyen TBH, Leonzio G, Galanopoulos C, Zondervan E (2018) Integration of carbon dioxide and hydrogen supply chains. Comp Aid Chem Eng 43:1413–1418. https://doi.org/10.1016/B978-0-444-64235-6.50247-3

Olajire AA (2010) CO2 capture and separation technologies for end-of-pipe applications – a review. Energy 35(6):2610–2628. https://doi.org/10.1016/j.energy.2010.02.030

Ozkır V, Baslıgil H (2013) Multi-objective optimization of closed-loop supply chains in uncertain environment. J Clean Prod 41:114–125. https://doi.org/10.1016/j.jclepro.2012.10.013

Patricio J, Angelis-Dimakis A, Castillo-Castillo A, Kalmykova Y, Rosado L (2017) Region prioritization for the development of carbon capture and utilization technologies. J CO2 Util 17:50–59. https://doi.org/10.1016/j.jcou.2016.10.002

Pettit SJ, Beresford AKC (2005) Emergency relief logistics: an evaluation of military, non-military and composite response models. Int J Logist 8(4):313–331. https://doi.org/10.1080/13675560500407325

Pires JCM, Martins FG, Alvim-Ferraz MCM, Simoes M (2011) Recent development on carbon capture and storage: an overview. Chem Eng Res Des 89:1446–1460. https://doi.org/10.1016/j.cherd.2011.01.028

Rahmani D, Mahoodian V (2017) Strategic and operational supply chain network design to reduce carbon emission considering reliability and robustness. J Clean Prod 149:607–620. https://doi.org/10.1016/j.jclepro.2017.02.068

Rahmawati SD, Hoda MF, Kuntadi A (2015) CO2 injection project analysis using application of integrated model and optimization. In: SPE Middle East Oil and Gas Show and Conference, MEOS, Proceedings, pp 1916–1926

Romanenko EV, (2014) Carbon Capture Utilization and Storage Market Dynamics: Matching CO2 Supply and Demand for Enhanced Oil Recovery, Thesis submitted in partial fulfillment of the requirements of Master of Philosophy in System Dynamics (Universitetet i Bergen, Università degli Studi di Palermo) and Master of Science in Business Administration (Radboud Universiteit Nijmegen). http://bora.uib.no/bitstream/handle/1956/8740/124158780.pdf?sequence=1&isAllowed=y

Rubin ES, Davison JE, Herzog HJ (2015) The cost of CO2 capture and storage. Int J Greenhouse Gas Control 40:378–400. https://doi.org/10.1016/j.ijggc.2015.05.018

Samanta A, Zhao A, Shimizu GKH, Sarkar P, Gupta R (2012) Post-combustion CO2 capture using solid sorbents: a review. Ind Eng Chem Res 51:1438–1463. https://doi.org/10.1021/ie200686q

Serpa J, Morbee J, Tzimas E (2011) Technical and economic characteristics of aCO2Transmission pipeline infrastructure. https://doi.org/10.2790/30861

Solunke RD, Veser G (2011) Integrating desulfurization with CO2-capture in chemical looping combustion. Fuel 90:608–617. https://doi.org/10.1016/j.fuel.2010.09.039

Song J, Zhang D (2012) Comprehensive review of Caprock-Sealing mechanisms for geologic carbon sequestration. Environ Sci Technol 47(1):9–22. https://doi.org/10.1021/es301610p

Song C, Liu Q, Ji N, Deng S, Zhao J, Li Y, Song Y, Lid H (2018) Alternative pathways for efficient CO2 capture by hybrid processes—a review. Renew Sust Energ Rev 82:215–231. https://doi.org/10.1016/j.rser.2017.09.040

Song C, Liua Q, Deng S, Li H, Kitamura Y (2019) Cryogenic-based CO2 capture technologies: state-of-the-art developments and current challenges. Renew Sust Energ Rev 101:265–278. https://doi.org/10.1016/j.rser.2018.11.018

Sun L, Chen W (2015) Pipeline networks for CCUS by static programming in the Chinese mainland. Qinghua Daxue Xuebao/J Tsinghua Univ 55:678–683

Sun L, Chen W (2017) Development and application of a multi-stage CCUS source–sink matching model. Appl Energy 185:1424–1432. https://doi.org/10.1016/j.apenergy.2016.01.009

Tapia JFD, Lee JY, Ooi REH, Foo DCY, Tan RR (2016a) Optimal CO2 allocation and scheduling in enhanced oil recovery (EOR) operations. Appl Energy 184:337–345. https://doi.org/10.1016/j.apenergy.2016.09.093

Tapia JFD, Lee JY, Ooi REH, Foo DCY, Tan R (2016b) Planning and scheduling of CO2 capture, utilization and storage (CCUS) operations as a strip packing problem. Process Saf Environ Prot 104:358–372. https://doi.org/10.1016/j.psep.2016.09.013

Tapia JFD, Lee YJ, Ooi REH, Foo DCY, Tan RR (2018) A review of optimization and decision-making models for the planning of CO2 capture, utilization and storage (CCUS) systems. Suitable Prod Consump 13:1–15. https://doi.org/10.1016/j.spc.2017.10.001

Thiruvenkatachari R, Su S, An H, Yu XX (2009) Post combustion CO2 capture by carbon fibre monolithic adsorbents. Prog Energy Combust 35:438–455. https://doi.org/10.1016/j.pecs.2009.05.003

Tonziello J, Vellini M (2011) Oxygen production technologies for IGCC power plants with CO2 capture. Energy Procedia 4:637–644. https://doi.org/10.1016/j.egypro.2011.01.099

Van Wassenhove LN (2006) Humanitarian aid logistics: supply chain management in high gear. J Oper Res Soc 57:475–489. https://www.jstor.org/stable/4102445

Vikara D, Shih CY, Lin SM, Guinan A, Grant T, Morgan D, Donald Remson US (2017 December) DOE's economic approaches and resources for evaluating the cost of implementing carbon capture, utilization, and storage (CCUS), J. Sustainable Energy Eng 5(4)

Wu Q, Lin QG, Wang XZ, Zhai MY (2015) An inexact optimization model for planning regional carbon capture: transportation and storage systems under uncertainty. Int J Greenhouse Gas Control 42:615–628. https://doi.org/10.1016/j.ijggc.2015.09.017

Yue D, You F (2015) Integration of geological sequestration and microalgae biofixation supply chains for better greenhouse gas emission abatement. Chem Eng Trans 45:487–492. https://doi.org/10.3303/CET1545082

Zhang X, Fan JL, Wei YM (2013a) Technology roadmap study on carbon capture, utilization and storage in China. Energy Policy 59:536–550. https://doi.org/10.1016/j.enpol.2013.04.005

Zhang Y, Sunarso J, Liu S, Wang R (2013b) Current status and development of membranes for CO2/CH4 separation: a review. Int J Greenhouse Gas Control 12:84–107. https://doi.org/10.1016/j.ijggc.2012.10.009

Zhang S, Liu L, Zhang L, Zhuang Y, Du J (2018) An optimization model for carbon capture utilization and storage supply chain: a case study in northeastern China. Appl Energy 231:194–206. https://doi.org/10.1016/j.apenergy.2018.09.129

Zhu L, Liu X (2015) Promoting the carbon removal in coal utilization-a benefit-risk analysis among full-chain carbon capture and utilization project. Energy Environ 26:6–7. https://journals.sagepub.com/doi/pdf/10.1260/0958-305X.26.6-7.1035

Chapter 4
Natural Carbon Sequestration by Forestry

Xolile G. Ncipha and Venkataraman Sivakumar

Abstract Carbon dioxide (CO_2) levels have been rapidly rising in the atmosphere following the beginning of industrialisation and reaching new highs in the last recent years. Global carbon cycle studies established that the interannual variability in observed atmospheric CO_2 growth rates is attributed mainly to interannual inconsistent changes in carbon sequestration by terrestrial ecosystems, instead to oceans or changes in anthropogenic emissions. This terrestrial ecosystems sink is mainly in the forests of the world, which are enormous repositories of both biomass and soil carbon. In 1997, the United Nations Framework Convention on Climate Change (UNFCC) at Kyoto agreed on a protocol with an aim of reducing anthropogenic CO_2 emissions in the atmosphere through terrestrial ecosystems CO_2 sequestration. This ignited increased interest and efforts from the science community since the beginning of the twenty-first century, to probe the potential of terrestrial ecosystems CO_2 assimilation in offsetting the anthropogenic CO_2 impact in the atmosphere and subsequently in the global climate; this also inspired investigations on the role of environment and climate variables on the land ecosystems carbon sequestration.

This study reviews the developments in the understanding of the role of the terrestrial biosphere mainly the forests in anthropogenic carbon sequestration. It first examines the influence of the environment and climate variables in the global carbon cycle processes of the terrestrial ecosystems, including the effects of climate extremes. Then it assimilates the knowledge in global carbon budget generated in the past few decades. Studies indicate that the combined effect of nitrogen deposition, high atmospheric CO_2, warm climate, increased diffuse radiation support the capability of the terrestrial ecosystems to assimilate atmospheric carbon. Extreme climate change events and disturbances (ECE&D) can transform land ecosystems from sinks into sources. The aggregate of the global terrestrial ecosystems CO_2 net sink has been growing in the sequence of 0.1 ± 0.8, 1.1 ± 0.9 and 1.5 ± 0.9 Pg C

X. G. Ncipha (✉)
South African Weather Service, Pretoria, South Africa

School of Chemistry and Physics, University of KwaZulu-Natal, Durban, South Africa
e-mail: xolile.ncipha@weathersa.co.za

V. Sivakumar
School of Chemistry and Physics, University of KwaZulu-Natal, Durban, South Africa

Inamuddin et al. (eds.), *Sustainable Agriculture Reviews 37*, Sustainable Agriculture Reviews 37, https://doi.org/10.1007/978-3-030-29298-0_4

year $^{-1}$, in the past few decades of the 1980s, 1990s and 2000s, respectively. Current global tree cover has increased by 7.1% relative to the 1982 level. The global forests are a substantial and persistence carbon sink.

Keywords Atmospheric carbon dioxide · Climate change · Sink · Source · Forests · Carbon dioxide-cycle · Nitrogen-cycle · Climate extremes and disturbance · Gross primary production · Total ecosystem respiration · Net ecosystem exchange

Abbreviations

CO_2	Carbon dioxide
F	Disturbance by fire
GPP	Gross Primary Production
H	Disturbance by harvest
IPCC	Intergovernmental Panel on Climate Change
NBP	Net Biome Production
NEE	Net Ecosystem Exchange
NEP	Net Ecosystem Production
NPP	Net Primary Production
R_a	Plant respiration
RE	Total ecosystem respiration
R_h	Heterotrophic respiration
WMO	World Meteorological Organisation

4.1 Introduction

Atmospheric carbon dioxide (CO_2) is the principal anthropogenic greenhouse gas in the atmosphere with a radiative forcing of about 1.66 W/m^2, contributing about 65% to radiative forcing by greenhouse gases (Forster et al. 2007; Churkina et al. 2009; World Meteorological Organisation (WMO) 2014). The greenhouse gases block heat in the atmosphere from escaping to space by absorbing infrared radiation emitted by the Earth's surface. Substantial evidence is already established, it reveals that greenhouse gases have warmed the global mean surface temperature by 0.3–0.6 °C over the last century. Recent general circulation models projections indicate that the average global temperature will rise between 0.3 °C and 4.8 °C by the end of this century, with larger warming at higher latitudes (Intergovernmental Panel on Climate Change (IPCC) 1996; Rustad et al. 2000; Frank et al. 2015). Among the greenhouse gases, CO_2 is responsible for about 84% of the growth in radiative forcing over the last decade (WMO 2012, 2013, 2014). The increasing CO_2 concentration in the atmosphere from anthropogenic emissions since the industrial revolution is viewed as the predominant human influence contributing to global climate change (IPCC 1996; Denman et al. 2007; Forster et al. 2007; Deutscher et al. 2014).

In 2012, levels of CO_2 in the atmosphere reached 392.52 ppm and Mauna Loa the station with the longest global records of CO_2 recorded maiden daily averages above 400 ppm in May 2013 (Le Quéré et al. 2014). The globally averaged CO_2 reached a new high in 2013 of 396 ppm (WMO 2014). Before the beginning of the industrial revolution in 1750, atmospheric CO_2 concentration ranged between 180 ppm and 290 ppm. These levels constituted a natural equilibrium of fluxes between the atmosphere, the oceans and the terrestrial biosphere (WMO 2012; Ciais et al. 2013; WMO 2013, 2014).

Table 4.1 shows the world's human activities based CO_2 budget, built-up since the start of the industrial period and averaged over the 1980s, 1990s, and 2000s. During the period 1750–2011, the fossil fuel burning and cement manufacturing emitted 375 ± 30 Pg C (1 Pg C = 10^{15} g C) to the atmosphere. Land use activities, chiefly forest removal, emitted an additional 180 ± 80 Pg C. From the total of 555 ± 85 Pg C of anthropogenic carbon released to the atmosphere from fossil fuel combustion, cement production, and land use change, only 240 ± 10 Pg C have accumulated and remained in the atmosphere. The other carbon from human activities based emissions has been taken up by the carbon sinks, the ocean, and land ecosystems. The ocean accumulated 155 ± 30 Pg C of anthropogenic carbon since industrialisation Era. The undisturbed terrestrial ecosystems have amassed 160 ± 90 Pg C of human activities based carbon since industrialisation, hence not fully compensating the net CO_2 losses to the atmosphere from terrestrial ecosystems from land use change approximated at 180 ± 80 Pg C during the same time. This undisturbed terrestrial ecosystems carbon storage is called the "residual sink" (Esser et al. 2010; Ciais et al. 2013); a phrase that covers land use change practices like forest regrowth caused by abandonment of agricultural land in mid-latitudes, as well as eco-physiological developments like strengthened forest growth ascribed to CO_2 and nitrogen fertilisation, and response to warming climate (Malhi et al. 1999).

This terrestrial "residual sink" is mainly in the forests of the world, which are enormous repositories of both biomass and soil carbon (Malhi et al. 1999). About 28% of global land surface is occupied by forests and they hold 77% of all terrestrial

Table 4.1 Global anthropogenic CO_2 balance accrued from the beginning of the Industrial Era in 1750 and its decadal averaged variation from the 1980s to 2000s. The ocean and land sinks are presented as negative values

	1750–2011 (Pg C)	1980–1989 (Pg C year $^{-1}$)	1990–1999 (Pg C year $^{-1}$)	2000–2009 (Pg C year $^{-1}$)
Atmospheric increase	240 ± 10	3.4 ± 0.2	3.1 ± 0.2	4.0 ± 0.2
Fossil fuel burning and cement manufacturing	375 ± 30	5.5 ± 0.4	6.4 ± 0.5	7.8 ± 0.6
Ocean to atmosphere flow	−155 ± 30	−2.0 ± 0.7	−2.2 ± 0.7	−2.3 ± 0.7
Land-to-atmosphere net carbon balance	30 ± 45	−0.1 ± 0.8	−1.1 ± 0.9	−1.5 ± 0.9
Net land-use change	180 ± 80	1.4 ± 0.8	1.5 ± 0.8	1.1 ± 0.8
Residual land sink	−160 ± 90	−1.5 ± 1.1	−2.6 ± 1.2	−2.6 ± 1.2

Modified after Ciais et al. (2013)

above-ground carbon (Baccini et al. 2012; Goodman and Herold 2014). Globally tropical forests have the largest spatial cover, most dense carbon density and are most diverse. Tropical forests make-up just over 50% of the world's remaining forests (Malhi and Grace 2000; Goodman and Herold 2014) and 22% of the global potential vegetation area (Tan et al. 2010). They are estimated to be responsible for 59% of global carbon stored in forests (Tan et al. 2010).

Forests role in the global carbon cycle is key as they exchange trace gases between the atmosphere and biosphere (Rodda et al. 2016) and they are substantial atmospheric carbon sinks (Malhi and Grace 2000; Saleska et al. 2003; Tan et al. 2010; Pan et al. 2011; Baccini et al. 2012; Vieilledent et al. 2013; Rodda et al. 2016). They are reported to assimilate more than a third of the total anthropogenic carbon in the atmosphere every year (Schulze et al. 1999; Myneni et al. 2001; Saleska et al. 2003; Erisman et al. 2011; Fernández-Martínez et al. 2018). The net carbon budget of a forest is determined by a delicate carbon fluxes equilibrium of carbon assimilation process known as photosynthesis, and processes of carbon release known as respiration and disturbance. These carbon flux processes function on a range of timeframes; from daily, seasonal, interannual, interdecadal and beyond. The global carbon cycle processes are impacted by climate and environment parameters, like temperature, water availability, and prevalence of disturbances (Malhi et al. 1999). The land biosphere carbon cycle processes can be classified as; gross primary production (GPP) which is carbon absorption through photosynthesis, net primary production ($NPP = GPP - R_a$) which is a proportion of gross primary production giving rise to growth when plant respiration (R_a) is taken into account, net ecosystem production or exchange (NEP or $NEE = GPP - R_a - R_h = GPP - RE$) which is a constituent of gross primary production leading to growth when plant respiration and heterotrophic respiration of soil organisms (R_h) is taken into consideration, and the net biome production ($NBP = GPP - R_a - R_h - F - H$) which is the plant net carbon flux which in addition considers non-respiratory carbon release like disturbance by fire (F) and harvest (H) (Schulze et al. 2000; Kutsch et al. 2010).

The global aggregate of the terrestrial ecosystems CO_2 net sink including the offsetting role of land use change has been growing in the sequence of 0.1 ± 0.8, 1.1 ± 0.9 and 1.5 ± 0.9 Pg C year $^{-1}$, in the past few decades of the 1980s, 1990s, and 2000s, respectively (Ciais et al. 2013). In recognition of the probable climatic impacts of rising atmospheric CO_2 levels. The Kyoto protocol was ratified by the United Nations Framework Convention on Climate Change (UNFCC) in 1997 with an aim of mitigating anthropogenic CO_2 in the atmosphere through terrestrial ecosystems CO_2 sequestration. The protocol commits industrialised party nations to use forest biomass sinks to fulfil their carbon emission mitigation obligations. Through conservation of forests carbon sinks, which include the establishment of new forests and restoration of existing forests (Schulze et al. 2000; Myneni et al. 2001). This chapter reviews the developments in understanding the role of the terrestrial biosphere mainly the forests in anthropogenic carbon sequestration. It first examines the influence of the environment and climate variables in the global carbon cycle processes of the terrestrial ecosystems, including the effects of climate extremes. Then it assimilates the knowledge in global carbon budget generated in the past few decades.

4.2 Influence of the Environment and Climate Variables in the Global Carbon Cycle

4.2.1 Nitrogen Fertilisation

Earlier studies acknowledged the important role played by the cooperation between carbon and nitrogen biogeochemical cycles within terrestrial ecosystems in influencing the long-term development of the plant, litter, and soil organic matter carbon and nitrogen reserves, including the feedback of these reserves to changes in atmospheric composition and climate. Model simulation studies have established that this cooperation will potentially impact the future course of atmospheric CO_2 concentration and associated climate changes, as the interactions of both cycles affect net CO_2 uptake from the atmosphere by terrestrial biosphere ecosystems (Thornton et al. 2009; Churkina et al. 2009; Erisman et al. 2011). The biogeochemical cycles of nitrogen and carbon are closely linked to each other owing to the nutritional requirements of plants for these two nutrients (Ciais et al. 2013). Nitrogen is an elemental constituent of plant proteins. Plant proteins play an important role in assembling carbohydrates during photosynthesis hence they are important for plant growth rate and CO_2 fixation. The source of this nitrogen is the reactive nitrogen compounds from natural sources such as lightning, wildfires and biological nitrogen gas (N_2) fixation and anthropogenic sources which are mainly fossil fuel combustion emissions (Erisman et al. 2011; Esser et al. 2010; Ciais et al. 2013).

Recent results from field studies indicated nitrogen accessibility as a limit to high CO_2 concentration stimulated vegetation development. The nitrogen deficiency in the soils seemed to stifle the favourable feedback of plant development to high CO_2. A rise in anthropogenic nitrogen deposition improves its accessibility in soil and this augments the terrestrial biosphere sink (Churkina et al. 2009; Erisman et al. 2011; Smith et al. 2015; Sakaguchi et al. 2016). The global warming climate is understood to accelerate the natural nitrogen-cycle, this process naturally increases the nitrogen accessibility to vegetation. The process works through more rapid litter decomposition and associated nitrogen mineralisation rates (Erisman et al. 2011). The increase in nitrogen accessibility in soil has been revealed to have a positive association with carbon assimilation by European forests. In spite of the global continuous increase in the amount of reactive nitrogen deposited mostly on land since the beginning industrialisation, nitrogen nutrient levels are still inadequate for global terrestrial ecosystems (Churkina et al. 2009). The global distribution of reactive nitrogen varies regionally due to regional differences in the intensity of anthropogenic sources of nitrogen. Many temperate and boreal ecosystems are nitrogen limited, unlike many tropical ecosystems with greater nitrogen availability (Hietz et al. 2011). Several global scale studies were undertaken in the last few decades investigating the response of terrestrial ecosystems to growing nitrogen deposition. Results from ^{15}N-tracer field experiments indicated that the elevated nitrogen levels in the atmosphere resulted in a modest impact to terrestrial carbon absorption with 0.25 Pg C year $^{-1}$ being further sequestered. In contrast, estimates based on various models showed a considerable increase in terrestrial carbon sequestration in nitrogen

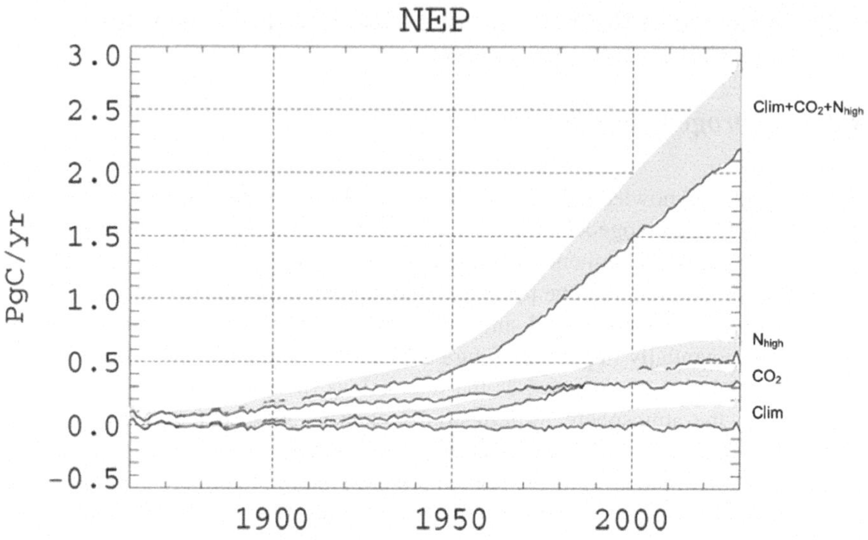

Fig. 4.1 The synergetic impacts of elevating atmospheric CO_2 and nitrogen deposition on carbon sequestration by terrestrial ecosystems in a warming global climate. Greyed out areas indicate the impact of plants on net ecosystem production (NEP) feedback. The top boundary of the greyed out area was simulated with prospective plant cover. The dark bottom boundary was simulated with current plant cover presumed to be carbon neutral. (Reprinted with permission of American Geophysical Union from Churkina et al. (2009))

enriched land. The carbon uptake ranged between 0.75 Pg C and 2.21 Pg C during the 1990s, this increase relied on the size and age of regrowing forests (Churkina et al. 2009).

Churkina et al. (2009) used a unified model of climate, ocean, and land biogeochemistry to probe the impact of elevating atmospheric deposition of nitrogen and CO_2 on the capacity of global terrestrial biosphere atmospheric CO_2 sequestration between the period 1860–2030. The results showed that there is a substantial combined impact on raising CO_2 and nitrogen deposition on the world's terrestrial ecosystems CO_2 sequestration (Fig. 4.1). During the 1990s, the worldwide synergistic impact on net ecosystem production was around 47%, which was greater than either the impact of raising nitrogen deposition exclusively which was 29% or CO_2 fertilisation effect which raised the net ecosystem production by 24%. Esser et al. (2010) performed similar modelling experiments and also found that the overall impact of the combination of CO_2 and nitrogen deposition on terrestrial biosphere CO_2 sequestration was considerably greater than the individual influences added together. This synergistic nonlinear effect on the terrestrial ecosystem's productivity is concomitantly limited by variables such as climate, CO_2 and nitrogen's presence. The model simulations results also found that the greatest growth in land ecosystems CO_2 uptake occurred in industrialised areas like Eastern North America, Europe, and South East Asia. In 2030, the regions with carbon sequestration flux growth greater than 0.2 Kg C m^{-2} year $^{-1}$ were not only the long-established industrialised areas of the world; developing industrial regions like Africa and South America are included. The strength of terrestrial ecosystems feedback to the synergistic impact of atmospheric CO_2 and nitrogen deposition, and its implication to atmospheric CO_2 levels and global temperature is dependent to biome types and their dispersion in the industrialised areas. The most intense land ecosystems response was obtained in expansive forest regions in Europe, North America, and Asia (Churkina et al. 2009).

4.2.2 Temperature and Soil Water Availability

The potential positive impact of high atmospheric CO_2 on plants growth could be influenced by the warming global climate and altered precipitation patterns. As water availability and temperature are both global strong drivers of photosynthesis and respiration (Fernández-Martínez et al. 2018). In forests ecosystems, soil respiration rates are mainly driven by temperature and to a smaller degree by the amount of water in the soil (Malhi et al. 1999). The influence of climate can in few instances change a land ecosystem from a carbon sink to a source or the other way around (Valentini et al. 2000). As the above-ground carbon dominates the forest biomass (Baccini et al. 2012; Goodman and Herold 2014), however, the total ecosystem respiration of forests is commanded by roots and microbial soil respiration of organic matter (Malhi et al. 1999; Valentini et al. 2000; Davidson and Janssens 2006; Bond-Lamberty and Thomson 2010). In forests in tropical regions, autotrophic and heterotrophic respiration processes within soil biomass account for over 70% of respiratory CO_2 (Malhi et al. 1999).

Soil respiration has a diurnal and seasonal variation which is influenced by the soil temperature and moisture. Diurnally it reaches its mean maximum during the night-time CO_2 release and its mean minimum during the daytime CO_2 uptake. It reaches its seasonal maximum during the summer, which is June to August in the Northern Hemisphere (Fig. 4.2) and December to February in the Southern Hemisphere. This seasonal variation is as a result of that the leading driving influence of respiration in boreal and temperate forests is the temperature, influences like drought stress are not commanding. The respiration reaches minimum levels in winter. The seasonality of soil respiration in tropical forests is less pronounced (Malhi et al. 1999; Saigusa et al. 2008; Mkhabela et al. 2009).

Valentini et al. (2000) ran automated eddy covariance observations of CO_2 fluxes over 15 forests in Europe covering approximately 40°N – 65°N latitudinal range between 1996 and 1998. They found a conservative change of gross primary production with latitude and the total respiration was growing stronger with the latitude over the study region. High latitude boreal soils hold a greater quantity of soil organic matter in an unstable form that is susceptible to swift decay than temperate soils. The effective temperature sensitivity (Q10) of soil organic matter decay is far greater in cooler than in warmer subtropical climates. Temperature rises in chilly areas influence decay rates greater than net primary productivity (Valentini et al. 2000). This increase of total respiration with latitude is explained by the fact

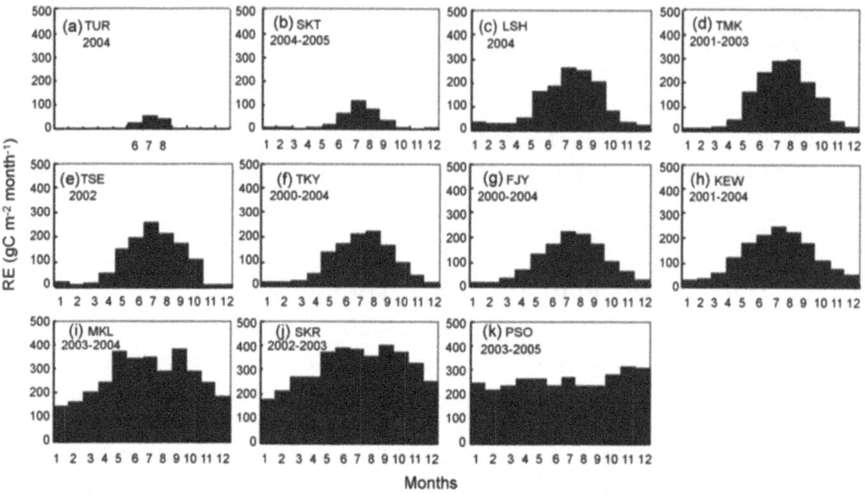

Fig. 4.2 Monthly total ecosystem respiration (RE) measured at the eleven locations in East Asia. The codes TUR, SKT, LSH, TMK, TSE, TKY, FJY, KEW, MKL, SKR and PSO represent different sites in the region and the years represent periods of RE analysis. (**a**–**b**) represent the boreal sites, (**c**–**h**) temperate sites and (**i**–**k**) tropical sites. (Reprinted with permission of Elsevier from Saigusa et al. (2008))

that northern high latitudes temperatures have risen by plus 4 °C, while southern high latitudes temperatures have increased less. The ratio net ecosystem exchange and total respiration (NEE/RE) grows with latitude in Fig. 4.3, showing that total ecosystem respiration (RE) grows larger for the cold northern high latitude sites and this clarifies the drop in net ecosystem exchange (Valentini et al. 2000). Warm winters are likely to transform old boreal forests sites from carbon sink to sources as they increase the annual amount of total respiration (Malhi et al. 1999; Valentini et al. 2000).

Roots respiration and microbial decay are also affected by soil moisture deficiency. As a result, most observations data dependent models link soil respiration to temperature and frequently to soil moisture or precipitation as well (Davidson and Janssens 2006). Lack of soil wetness can affect soil carbon and nutrients release to plants by limiting microbial decay. In the tropical regions, the soil wetness deficit effects may be severe in a dry season, as a result of excessive evapotranspiration rates and low available moisture content in many old, severely eroded tropical clay soils (Malhi et al. 1999). The evidence of the important role of soil water availability to terrestrial carbon cycle processes including respiration was presented during the Europe heat and drought which occurred in 2003; it caused a reduction in primary productivity and respiration. During this heat wave and drought, it was observed the gross primary production and total respiration anomalously dropped (Fig. 4.4). The Europe-wide model simulated gross primary production anomalies during 2003 correlated more with precipitation variations than with summer air temperature. This indicated the commanding part of soil moisture deficiencies to the plant

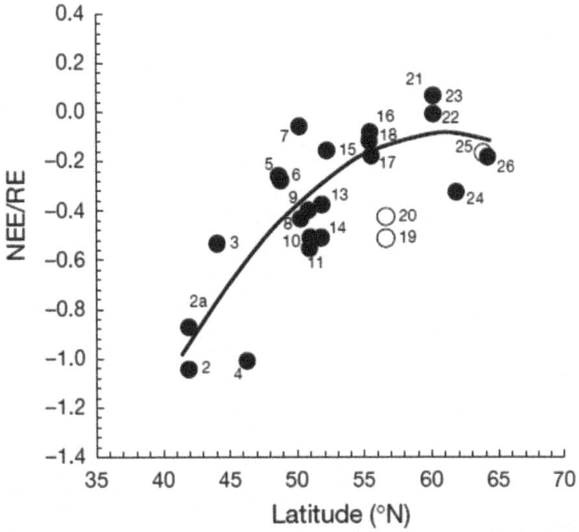

Fig. 4.3 The ratio of net ecosystem exchange and total ecosystem respiration (NEE/RE) against northern hemisphere latitude. A minus sign represent carbon absorption by the forest, whereas plus sign represent carbon release to the atmosphere by the forest. (Reprinted with permission of Springer Nature from Valentini et al. (2000))

production processes. The reduction in plant and soil respiration was due to diminished plant photosynthetically produced substrates and microbial soil respiration which was caused by severe dry conditions (Ciais et al. 2005).

4.2.3 Radiation

The sunshine incident on the surface of the Earth is the main regulator of plant photosynthesis. Photosynthesis on leaf surface grows nonlinearly with incident photosynthetically active radiation (Kobayashi et al. 2005; Mercado et al. 2009). Previous studies have identified that fluctuation in photosynthetically active radiation owing to aerosols from urban areas, volcanoes, biomass smoke, and cloud cover changes have a considerable effect on terrestrial ecosystems CO_2 uptake (Kobayashi et al. 2005). Aerosols particles scatter sunlight back into space and hence reduce the solar radiation reaching the Earth's surface, a phenomenon called 'global dimming'. This results in cooling of temperatures at the surface. Cooling has been observed following major volcanic eruptions over the last century, however, it was hard to quantify (Jones and Cox 2001). The eruption of Mount Pinatubo in the Philippines in June 1991 released huge amounts of aerosol particles into the stratosphere, which spread around the world within a month. This eruption was followed by a drop in the observed global surface temperatures, which were cooler than the 1958–1991 average, and cooler than 1991 alone. Model simulations showed that the surface cooling led to a reduction to soil and plant respiration globally, and a substantial increase in net carbon absorption of 1–2 Pg C year $^{-1}$ by land biosphere for several years after the eruption (Jones and Cox 2001; Lucht et al. 2002; Mercado et al. 2009).

Kobayashi et al. (2005) investigated the effect of reduction of the photosynthetically active radiation by heavy smoke from forest fires in Southeast Asia. Their

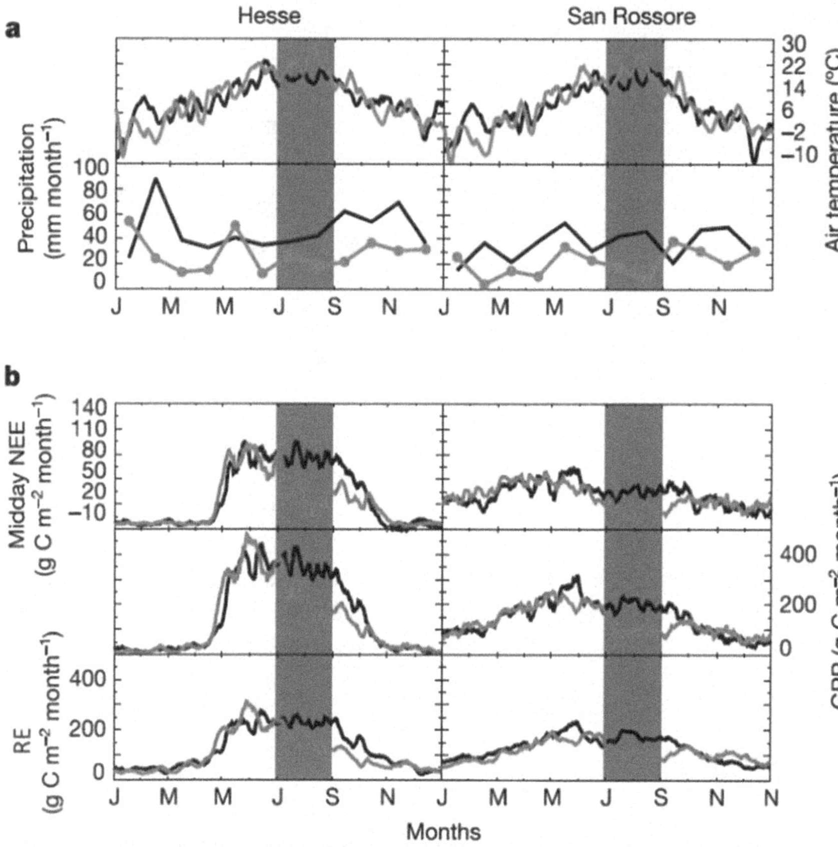

Fig. 4.4 Measured precipitation, temperature, gross primary production (GPP), net ecosystem exchange (NEE) and total ecosystem respiration (RE) fluxes during 2002 and 2003 at Hesse in France and San Rossore in northern Italy forest sites. (**a**) is climate data and (**b**) are carbon fluxes, data for 2002 are in black and for 2003 in colour. The northern hemisphere summer season July to August is greyed out. Positive NEE represents carbon absorption and negative represents carbon release. (Reprinted with permission of Springer Nature from Ciais et al. (2005))

model simulations indicated that the reduction of photosynthetically active radiation by smoke reduced the net primary production over the affected areas. Mercado et al. (2009) argued that, the net effect of radiation variations linked with a rise in clouds or aerosols scattering on photosynthesis relies on a balance between the depletion in total photosynthetically active radiation, which leads to reduction in photosynthesis (Kobayashi et al. 2005); and the growth in the diffuse portion of the photosynthetically active radiation, which leads to a rise in photosynthesis. Though some global climate/carbon cycle models incorporate the impacts of atmospheric particulate matter on total irradiance and surface temperature, some don't account for the impacts of clouds and aerosols on the terrestrial carbon assimilation through variations in the diffuse fraction radiation. Mercado et al. (2009) investigated the effect of

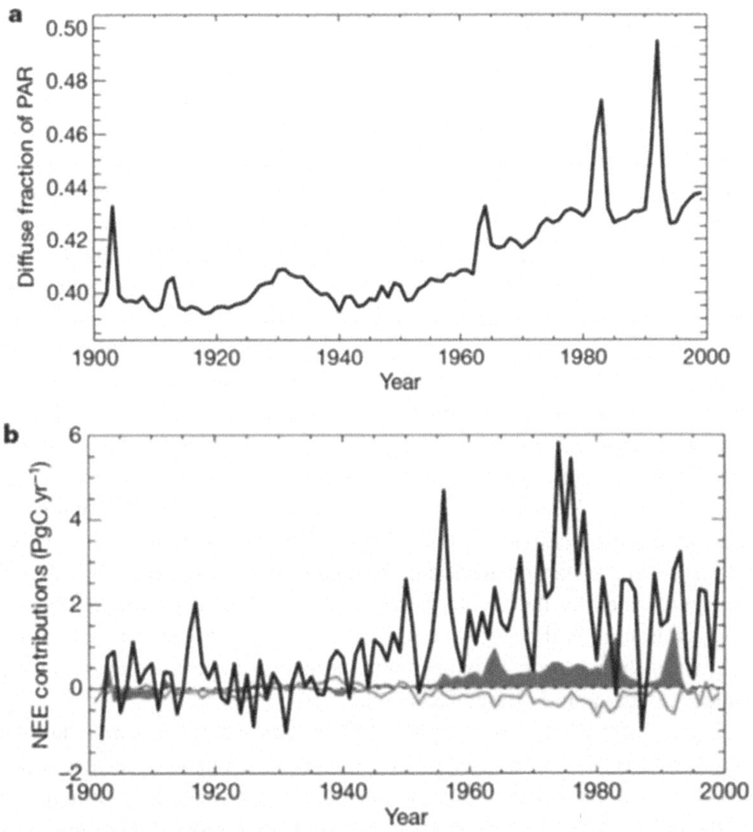

Fig. 4.5 The effect of variations in the scattered fraction of solar radiation on the land carbon absorption during the twentieth century. (**a**) HadGEM2-A model output of a global average of a yearly diffuse portion of the photosynthetically active radiation, based on aerosol optical thickness from volcanic and anthropogenic sources. (**b**) Simulated land net ecosystem exchange (NEE) (red) under the influence of simulated diffuse fraction of solar radiation, determined from the difference between simulated NEE under the control of changing diffuse fraction radiation (black, total NEE) and simulated NEE (green) under the influence of the total photosynthetically active radiation. Here a positive NEE represent carbon absorption, whereas a negative NEE represent carbon release. (Reprinted with permission of Springer Nature from Mercado et al. (2009))

variation of scattered radiation on the global terrestrial biosphere carbon cycle. Their model simulations results indicated the increase of 0.44 Pg C year $^{-1}$ of land carbon sink, during the dimming period (1960–1980) associated with increased anthropogenic solar light attenuation (Fig. 4.5). Their results were in agreement with Jones and Cox (2001) and Lucht et al. (2002) results on the effect of Pinatubo eruption impact on global carbon cycle when the effect of increased diffuse radiation is considered, the results were also in agreement with Kobayashi et al. (2005) results on the impact of solar attenuation by aerosols on land biosphere carbon uptake when the changes in diffuse light are not taken into account.

4.2.4 Climate Extremes and Disturbance

A climate extreme is described as the incidence or occurrence of a climate parameter value outside the range of its observed values in a specific region within a defined climate reference period (Frank et al. 2015). Considerable changes occurring in between years in recently observed CO_2 increase rates in the atmosphere arise mainly from variations in carbon sequestration by terrestrial ecosystems, instead of oceans or fluctuations in anthropogenic emissions (Jung et al. 2017). Extreme climate events and disturbances are believed to be the primary sources of the interannual variations of the land biosphere carbon fluxes which are often pronounced (Xiao et al. 2016), and in return, they influence the atmospheric CO_2 interannual variations (Jung et al. 2017). Extreme climate events such as heavy precipitation, heat waves and droughts; and disturbances such as fire, hurricanes, cyclones, wind storms, and insect flare-ups can considerably change the land's biosphere make-up and operation, and impact on land carbon flux exchanges (Frank et al. 2015; Xiao et al. 2016). The science community interest in the influences of extreme climate events and disturbances on terrestrial carbon cycle increased as from the start of the twenty-first century (Xiao et al. 2016). This might have been motivated by the Kyoto protocol of the UNFCC on reducing anthropogenic CO_2 emissions in the atmosphere through terrestrial ecosystems CO_2 sequestration. Most of the research activities on the impacts of extreme climate events and disturbances are case study based (Xiao et al. 2016).

The decade 2000–2009 was recorded as the warmest since instrumental observations began and it was associated with widespread drought occurrence globally. Throughout this decade several regions in the world experienced droughts which reduced the net primary production in those respective regions. Over this decade the Southern Hemisphere net primary production decreased, this countered the raised net primary production over the Northern Hemisphere (Zhao and Running 2010). The Europe-wide heat and drought which occurred in 2003 were estimated to have resulted in a 30% decline in gross primary production over Europe in 2003. A strong anomalous resultant loss of CO_2 of 0.5 Pg C year $^{-1}$ released to the atmosphere occurred, the loss had a reversal consequence of 4 years of net ecosystem carbon assimilation accumulation. It was found the gross primary production decreased together with respiration, instead of accelerating with the temperature rise. This heat wave and drought caused an unprecedented reduction in gross primary production in Europe in the last century (Ciais et al. 2005). Globally the drought during this decade was estimated to have reduced the global net primary production by 0.55 Pg C year $^{-1}$ (Zhao and Running 2010).

Heavy precipitation occurrences can change soil CO_2 fluxes and CO_2 assimilation by vegetation during periods the surface is water saturated. It may cause flood-related vegetation destruction and may cause nutrient loss through topsoil erosion, with losses of nutrients in a form of solid particulate and dissolved organic carbon from land and lotic ecosystems. Ice storms a type of extreme precipitation, the precipitation freezes soon after hitting the land surface. The swelling layer of ice can accumulate into a considerable weight on vegetation and cause damage to branches, limbs, or uproot the entire tree (Frank et al. 2015).

Several investigations have revealed that augmented precipitation supports plant growth of temperate biomes (Fang et al. 2005). Fang et al. (2005) determined that the feedback of vegetation to excessive precipitation also depends on the biome type. The net primary production of the temperate grassland and deciduous broadleaf forest of China grew with precipitation. However, the cold-temperate deciduous-coniferous forests production decreased with increased precipitation (Fang et al. 2005). Extreme precipitation and seasonal changes in precipitation are likely to increase above-ground net primary production of dry biomes and reduce one of the mesic biomes (Xiao et al. 2016).

Extreme wind storms and tropical cyclones can cause huge perturbation to terrestrial ecosystems carbon cycle. They can result in tree mortality through the removal of leaves, branches injury or destruction, complete tree uprooting and water submerging by storm gushes. In forests, tree uprooting can result in an enduring effect on the carbon balance; through complete trees destruction and dry dead tree branches build-up that enables insect infestation or enormous blazes. Separate intense storms and cyclones can substantially affect the carbon balance of a large territory. In October 2005, Hurricane Wilma caused huge destruction over the Yucatán peninsula with exceptionally strong winds. A sudden reduction in leaf area and yield were detected (Frank et al. 2015).

Combustion associated reduction or complete exhaustion of biomass or soil organic matter is largely the outcome of climate extremes. The regularity and strength of fire are highly dependent to climate extremes such as precipitation, relative humidity, air temperature and wind (Frank et al. 2015). Extreme fire events release directly and immediately huge amounts of carbon to the atmosphere. The subsequent effect of biomass fires is changes in net ecosystem production, as a consequence of the reduction of gross primary production and ecosystem respiration of the remaining live stand and heterotrophic respiration of damaged biomass. Fire events may have an enduring consequence on vegetation make-up, soil constituent's arrangements, and soil hydrophobicity and mineral nutrient availability (Frank et al. 2015; Xiao et al. 2016).

There is a general agreement in the face of numerous uncertainties that climate conditions have an impact on the intensity and frequency of insect and pathogen flare-up, through variations in distribution, breeding, the evolution of the host plant, and mortality and dispersal span changes of insect herbivores. Distinct kinds of climate extremes may then speed up insect and pathogen flare-ups, resulting in an indirect and delayed impact on the terrestrial ecosystems carbon cycle. Rapid insect population growth appears to be favoured by warm temperatures, as the mortality is reduced in the cold season and accelerated insect development rates in warm conditions. There are several examples of the impact of insects and pathogens on the terrestrial ecosystem (Frank et al. 2015). Like the unprecedented herbivore outbreak over millions of hectares of coniferous forests of western North America, that occurred over past decades. It resulted in tree mortality to a section of the global tree owing to a synergistic effect of high temperatures, dry conditions, and concomitant herbivory. The impact of this disruption at the site scale is minor ecosystem carbon reduction; however, the loss was huge at a regional scale (Xiao et al. 2016).

4.3 Forests Global Carbon Sink

As interannual variations in observed atmospheric CO_2 growth rates is attributed mainly to interannual fluctuations in carbon sequestration by terrestrial ecosystems, instead to oceans or variations in anthropogenic emissions (Bousquet et al. 2000; Jung et al. 2017). There have been studies in quantifying the global carbon flux exchanges and repository in forests ecosystems in the last 3 decades. These studies were based in integrating field and satellite observations together with coupled models systems (Bousquet et al. 2000; Myneni et al. 2001; Beer et al. 2010; Jung et al. 2011), and bottom-up forest inventory methods (Pan et al. 2011).

Forests are huge reservoirs of terrestrial above and underground carbon (Baccini et al. 2012; Goodman and Herold 2014). The recent bottom-up carbon stock in the global forests is estimated to be 861 $\pm$ 66 Pg C, with 383 $\pm$ 30 Pg C (44%) stored up to 1 m underground, 363 $\pm$ 28 Pg C (42%) in live above and underground biomass, 73 $\pm$ 6 Pg C (8%) in deadwood, and 43 $\pm$ 3 Pg C (5%) in debris. The tropical forests are the largest terrestrial carbon depository storing 471 $\pm$ 93 Pg C (55%), followed by boreal 272 $\pm$ 23 Pg C (32%), and then temperate 119 $\pm$ 6 Pg C (14%). The tropical and boreal forests have comparable carbon stock densities of 242 and 239 Mg C ha^{-1}, whilst the density in temperate forests is 155 Mg C ha^{-1}. Though tropical and boreal forests have comparable carbon stock densities, there is a contrast in the ecosystems structural carbon distribution storage. Tropical forests store 56% of the carbon in biomass and 32% in soil, while boreal forests keep only 20% in biomass and 60% in soil (Pan et al. 2011).

Photosynthesis a carbon assimilation process by plants, also known at the ecosystem level as gross primary production, is the biggest global carbon flux. It controls ecosystem functions like respiration and growth (Beer et al. 2010). Recent observations rooted in multi-model analysis estimate the global terrestrial gross primary production in the range 119–123 Pg C year $^{-1}$ (Beer et al. 2010; Jung et al. 2011). Table 4.2 shows the gross primary production for different biomes of the world. The tropical region biomes collectively dominate the global carbon assimilation with 60% contribution; the tropical forests contribute 34% and savannahs 26% to global terrestrial gross primary production. Figure 4.6 shows the global spatial distribution of the annual mean gross primary production in the last 3 decades. The gross primary production in the tropics is generally high throughout the year and has little seasonality. The boreal and temperate forests have a peak in gross primary production during the summer season (Saigusa et al. 2008).

Table 4.3 shows the global forest carbon budget over the period 1990–1999 and 2000–2007. The table also has classified the global forests by their climatic regions. Ciais et al. (2013) reported large carbon uptakes which did not take into account the impacts of land use, of 1.5 $\pm$ 1.1, 2.6 $\pm$ 1.2 and 2.6 $\pm$ 1.2 Pg C year $^{-1}$ for the 1980s,

Table 4.2 Gross primary production (GPP) for different ecological regions of global Earth

Biome	GPP (Pg C year $^{-1}$)	GPP = 2 × NPP (Pg C year $^{-1}$)
Tropical forests	40.8	43.8
Temperate forests	9.9	16.2
Boreal forests	8.3	5.2
Tropical savannahs and grasslands	31.3	29.8
Temperate grasslands and shrublands	8.5	14
Deserts	6.4	7
Tundra	1.6	1
Croplands	14.8	8.2
Total	121.7	125.2

Modified after Beer et al. (2010)

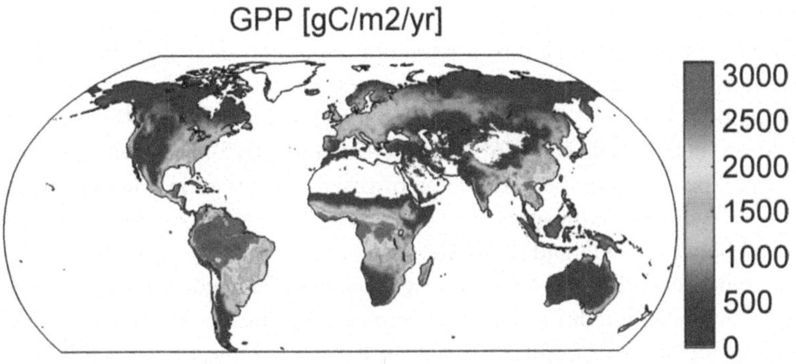

Fig. 4.6 Mean annual gross primary production (1982–2008) based on global observational upscaling of FLUXNET data. (Reprinted with permission of American Geophysical Union from Jung et al. (2011))

1990s, and 2000s, respectively. These carbon uptakes were slightly higher than 2.5 ± 0.4 Pg C year $^{-1}$ and 2.3 ± 0.5 Pg C year $^{-1}$ reported for the periods 1990–1999 and 2000–2007 in Table 4.3. The current assessment of global terrestrial carbon sink is 3.61 Pg C year $^{-1}$ for the decade 2007–2016 (Keenan and Williams 2018). The global net sink including the impacts of land use change of 0.1 ± 0.8, 1.1 ± 0.9 and 1.5 ± 0.9 Pg C year $^{-1}$ for the period 1980s, 1990s and 2000s, respectively (Ciais et al. 2013), is slightly higher than the one in Table 4.3. However, both global estimates of the net sink for the periods 1990s and 2000s show an increase from the 1990s to 2000s, indicating a substantial and persistence carbon sink in global forests (Pan et al. 2011; Ciais et al. 2013). The global average sink for boreal forests for the periods 1990–1999 and 2000–2007 remained the same, it slightly increased for temperate forests and decreased for tropical forests (Table 4.3). Aboveground carbon measurements were performed at the global tropical forests during the period 2003–2014. It was established that the world's tropical forests were a net source of carbon of 425.2 ± 31.0 Tg C year $^{-1}$ due to deforestation and forest degradation (Baccini et al. 2017).

Table 4.3 Global forest carbon balance for two recent time periods. Positive sign represent sinks and negative sources

Carbon sink and source in biomes	1990–1999 (Pg C year $^{-1}$)	2000–2007 (Pg C year $^{-1}$)
Boreal forest	0.50 ± 0.08	0.50 ± 0.08
Temperate forest	0.67 ± 0.08	0.78 ± 0.09
Tropical intact forest	1.33 ± 0.35	1.02 ± 0.47
Total sink in global forests	2.50 ± 0.36	2.30 ± 0.49
Tropical regrowth forest	1.57 ± 0.50	1.72 ± 0.54
Tropical gross deforestation emission	−3.03 ± 0.49	−2.82 ± 0.45
Tropical land-use change emission	−1.46 ± 0.70	−1.10 ± 0.70
Global gross forest sink	4.07 ± 0.62	4.02 ± 0.73
Global net forest sink	1.04 ± 0.79	1.20 ± 0.85

Modified after Pan et al. (2011)

Several studies were undertaken which present analysis of carbon dynamics for specific localities of the world, these studies report an increase in terrestrial sinks at these locations (Tian et al. 1998; Myneni et al. 2001; Fang et al. 2003; Saigusa et al. 2008; Amiro et al. 2010; Tan et al. 2010; Vilén et al. 2015). However, the forests in Europe may be showing the early signs of reaching saturation with regard to their capacity to assimilate atmospheric CO_2 further. The bottom-up forests inventories conducted between 2005 and 2010 are showing a decline in the total stem volume increment; the second sign is the recent decreasing trend in net forest area expansion, caused by deforestation due to land-use changes; the third indicator relates to the old forests growing stock susceptibility to disturbances, such as fire, storms and insects infestations (Nabuurs et al. 2013). Vilén et al. (2015) reported forest management interventions which increase the volume of growing stock in forests played a dominant role in the continued increase in the forest biomass carbon stock in European forests. Terrestrial ecosystems carbon sequestration holds much hope in the abatement of the anthropogenic global warming climate change. In contrast to the prevalent opinion that global forests coverage has diminished. The global tree cover has grown by 7.1% as compared to the 1982 coverage. This net growth in forests cover is the consequence of the net reduction in the tropical regions being exceeded by the net growth in the extratropical regions. Moreover, the global unvegetated land declined by 3.1% over the same time (Song et al. 2018).

References

Amiro BD, Barr AG, Barr JG, Black TA, Bracho R, Brown M, Chen J, Clark KL, Davis KJ, Desai AR, Dore S, Engel V, Fuentes JD, Goldstein AH, Goulden ML, Kolb TE, Lavigne MB, Law BE, Margolis HA, Martin T, McCaughey JH, Misson L, Montes-Helu M, Noormets A, Randerson JT, Starr G, Xiao J (2010) Ecosystem carbon dioxide fluxes after disturbance in forests of North America. J Geophys Res 115(4):G00K02. https://doi.org/10.1029/2010JG001390

Baccini A, Goetz SJ, Walker WS, Laporte NT, Sun M, Sulla-Menashe D, Hackler J, Beck PSA, Dubayah R, Friedl MA, Samanta S, Houghton RA (2012) Estimated carbon dioxide emissions from tropical deforestation improved by carbon-density maps. Nat Clim Chang 2:182–185. https://doi.org/10.1038/NCLIMATE1354

Baccini A, Walker W, Carvalho L, Farina M, Sulla-Menashe D, Houghton RA (2017) Tropical forests are a net carbon source based on aboveground measurements of gain and loss. Science 358(6360):230–234. https://doi.org/10.1126/science.aam5962

Beer C, Reichstein M, Tomelleri E, Ciais P, Jung M, Carvalhais N, Rödenbeck C, Arain MA, Baldocchi D, Bonan GB, Bondeau A, Cescatti A, Lasslop G, Lindroth A, Lomas M, Luyssaert S, Margolis H, Oleson KW, Roupsard O, Veenendaal E, Viovy N, Williams C, Woodward FI, Papale D (2010) Terrestrial Gross carbon dioxide uptake: global distribution and covariation with climate. Science 329:834–838

Bond-Lamberty B, Thomson A (2010) Temperature-associated increases in the global soil respiration record. Nature 464:579–583. https://doi.org/10.1038/nature08930

Bousquet P, Peylin P, Ciais P, Le Quéré C, Fiedlingstein P, Tans PP (2000) Regional changes in carbon dioxide fluxes of land and oceans since 1980. Science 290:1342–1346

Churkina G, Brovkin V, von Bloh W, Trusilova K, Jung M, Dentener F (2009) Synergy of rising nitrogen depositions and atmospheric CO_2 on land carbon uptake moderately offsets global warming. Global Biogeochem Cycles 23:GB4027. https://doi.org/10.1029/2008GB003291

Ciais P, Reichstein M, Viovy N, Granier A, Ogée J, Allard V, Aubinet M, Buchmann N, Bernhofer C, Carrara A, Chevallier F, De Noblet N, Friend AD, Friedlingstein P, Grünwald T, Heinesch B, Keronen P, Knohl A, Krinner G, Loustau D, Manca G, Matteucci G, Miglietta F, Ourcival JM, Papale D, Pilegaard K, Rambal S, Seufert G, Soussana JF, Sanz MJ, Schulze ED, Vesala T, Valentini R (2005) Europe-wide reduction in primary productivity caused by the heat and drought in 2003. Nature 437:529–533. https://doi.org/10.1038/nature03972

Ciais P, Sabine C, Bala G, Bopp L, Brovkin V, Canadell J, Chhabra A, DeFries R, Galloway J, Heimann M, Jones C, Le Quéré C, Myneni RB, Piao S, Thornton P (2013) Carbon and other biogeochemical cycles. In: Stocker TF, Qin D, Plattner G-K, Tignor M, Allen SK, Boschung J, Nauels A, Xia Y, Bex V, Midgley PM (eds) Climate change 2013: the physical science basis. Contribution of Working Group I to the Fifth Assessment Report of the Intergovernmental Panel on Climate. Cambridge University Press, Cambridge/New York

Davidson EA, Janssens A (2006) Temperature sensitivity of soil carbon decomposition and feedbacks to climate change. Nature 440:165–173. https://doi.org/10.1038/nature04514

Denman KL, Brasseur G, Chidthaisong A, Ciais P, Cox PM, Dickinson RE, Hauglustaine D, Heinze C, Holland E, Jacob D, Lohmann U, Ramachandran S, da Silva Dias PL, Wofsy SC, Zhang X (2007) Couplings between changes in the climate system and biogeochemistry. In: Solomon S, Qin D, Manning M, Chen Z, Marquis M, Averyt KB, Tignor M, Miller HL (eds) Climate change 2007: the physical science basis. Contribution of Working Group I to the Fourth Assessment Report of the Intergovernmental Panel on Climate Change. Cambridge University Press, Cambridge/New York

Deutscher NM, Sherlock V, Mikaloff Fletcher SE, Griffith DWT, Notholt J, Macatangay R, Connor BJ, Robinson J, Shiona H, Velazco VA, Wang Y, Wennberg PO, Wunch D (2014) Drivers of column-average CO_2 variability at southern hemispheric total carbon column observing network sites. Atmos Chem Phys 14:9883–9901. https://doi.org/10.5194/acp-14-9883-2014

Erisman JW, Galloway J, Seitzinger S, Bleeker A, Butterbach-Bahl K (2011) Reactive nitrogen in the environment and its effect on climate change. Curr Opin Environ Sustain 3:281–290. https://doi.org/10.1016/j.cosust.2011.08.012

Esser G, Kattge J, Sakalli A (2010) Feedback of carbon and nitrogen cycles enhances carbon sequestration in the terrestrial biosphere. Glob Chang Biol 17:819–842. https://doi.org/10.1111/j.1365-2486.2010.02261.x

Fang J, Piao S, Field CB, Yude Pan Y, Guo Q, Zhou L, Peng C, Tao S (2003) Increasing net primary production in China from 1982 to 1999. Front Ecol Environ 1(6):293–297

Fang J, Piao S, Zhou L, He J, Wei F, Myneni RB, Tucker CJ, Tan K (2005) Precipitation patterns alter growth of temperate vegetation. Geophys Res Lett 32:L21411. https://doi.org/10.1029/2005GL024231

Fernández-Martínez M, Sardans J, Chevallier F, Ciais P, Obersteiner M, Vicca S, Canadell JG, Bastos A, Friedlingstein P, Sitch S, Piao SL, Janssens IA, Peñuelas J (2018) Global trends in carbon sinks and their relationships with CO_2 and temperature. Nat Clim Chang 9:73–79. https://doi.org/10.1038/s41558-018-0367-7

Forster P, Ramaswamy V, Artaxo P, Berntsen T, Betts R, Fahey DW, Haywood J, Lean J, Lowe DC, Myhre G, Nganga J, Prinn R, Raga G, Schulz M, Van Dorland R (2007) Changes in atmospheric constituents and in radiative forcing. In: Solomon S, Qin D, Manning M, Chen Z, Marquis M, Averyt KB, Tignor M, Miller HL (eds) Climate change 2007: the physical science basis. Contribution of Working Group I to the Fourth Assessment Report of the Intergovernmental Panel on Climate Change. Cambridge University Press, Cambridge/New York, pp 129–234

Frank D, Reichstein M, Bahn M, Thonicke K, Frank D, Mahecha MD, Smith P, van der Velde M, Vicca S, Babst F, Beer C, Buchmann N, Canadell JG, Ciais P, Cramer W, Ibrom A, Miglietta F, Poulter B, Rammig A, Seneviratne SI, Walz A, Wattenbach M, Zavala MA, Zscheischler J (2015) Effects of climate extremes on the terrestrial carbon cycle: concepts, processes and potential future impacts. Glob Chang Biol 21:2861–2880. https://doi.org/10.1111/gcb.12916

Goodman RC, Herold M (2014) Why maintaining tropical forests is essential and urgent for a stable climate. Centre for Global Development, Washington, DC. CGD working paper 385. https://www.cgdev.org/publication/why-maintaining-tropical-forests-essential-and-urgent-stable-climate-working-paper-385

Hietz P, Turner BL, Wanek W, Richter A, Nock CA, Wright SJ (2011) Long-term change in the nitrogen cycle of tropical forests. Science 334:664–666. https://doi.org/10.1126/science.1211979

Intergovernmental Panel on Climate Change (1996) Climate change 1995: the science of climate change, contribution of Working Group I to the Second Assessment Report of the Intergovernmental Panel on climate change (IPCC). Cambridge University Press, Cambridge/New York, pp 1–572

Jones CD, Cox PM (2001) Modeling the volcanic signal in the atmospheric CO_2 record. Global Biogeochem Cycles 15(2):453–465

Jung M, Reichstein M, Margolis HA, Cescatti A, Richardson AD, Arain MA, Arneth A, Bernhofer C, Bonal D, Chen J, Gianelle D, Gobron N, Kiely G, Kutsch W, Lasslop G, Law BE, Lindroth A, Merbold L, Montagnani L, Eddy J, Moors EJ, Papale D, Sottocornola M, Francesco Vaccari F, Williams C (2011) Global patterns of land-atmosphere fluxes of carbon dioxide, latent heat, and sensible heat derived from eddy covariance, satellite, and meteorological observations. J Geophys Res 116:G00J07. https://doi.org/10.1029/2010JG001566

Jung M, Reichstein M, Schwalm CR, Huntingford C, Sitch S, Ahlström A, Arneth A, Camps-Valls G, Ciais P, Friedlingstein P, Gans F, Ichii K, Jain AK, Kato E, Papale D, Ben Poulter B, Raduly B, Rödenbeck C, Tramontana G, Viovy N, Wang Y-P, Weber U, Zaehle S, Zeng N (2017) Compensatory water effects link yearly global land CO_2 sink changes to temperature. Nature 000:1–11. https://doi.org/10.1038/nature20780

Keenan TF, Williams CA (2018) The terrestrial carbon sink. Ann Rev Environ Resour 43:1–25. https://doi.org/10.1146/annurev-environ-102017-030204

Kobayashi H, Tsuneo Matsunaga T, Hoyano A (2005) Net primary production in Southeast Asia following a large reduction in photosynthetically active radiation owing to smoke. Geophys Res Lett 32:L02403. https://doi.org/10.1029/2004GL021704

Kutsch WL, Aubinet M, Buchmann N, Smith P, Osborne B, Eugster W, Wattenbach M, Schrumpf M, Schulze ED, Tomelleri E, Ceschia E, Bernhofer C, Béziat P, Carrara A, Tommasi PD, Grünwald T, Jones M, Magliulo V, Marloie O, Moureaux C, Olioso A, Sanz MJ, Saunders M, Søgaard H, Ziegler W (2010) The net biome production of full crop rotations in Europe. Agric Ecosyst Environ 139:336–345. https://doi.org/10.1016/j.agee.2010.07.016

Le Quéré C, Peters GP, Andres RJ, Andrew RM, Boden TA, Ciais P, Friedlingstein P, Houghton RA, Marland G, Moriarty R, Stich S, Tans P, Arneth A, Arvanitis A, Bakker DCE, Bopp L, Canadell JG, Chini LP, Doney SC, Harper A, Harris I, House JI, Jain AK, Jones SD, Kato E, Keeling RF, Goldewijk KK, Körtzinger A, Koven C, Lefèvre N, Maignan F, Omar A, Ono T, Park G-H, Pfeil B, Poulter B, Raupach MR, Regnier P, Rödenbeck C, Saito S, Schwinger J, Segschneider J, Stocker BD, Takahashi T, Tilbrook B, van Heuven S, Viovy N, Wanninkhof R, Wiltshire A, Zaehle S (2014) Global carbon budget 2013. Earth Syst Sci Data 6:235–263. https://doi.org/10.5194/essd-6-235-2014

Lucht W, Prentice IC, Myneni RB, Sitch S, Friedlingstein P, Cramer W, Bousquet P, Buermann W, Smith B (2002) Climatic control of the high-latitude vegetation greening trend and Pinatubo effect. Science 296:1687–1689

Malhi Y, Grace J (2000) Tropical forests and atmospheric carbon dioxide. Trees 15(8):332–337

Malhi Y, Baldocchi DD, Jarvis PG (1999) The carbon balance of tropical, temperate and boreal forests. Plant Cell Environ 22:715–740

Mercado LM, Bellouin N, Sitch S, Boucher O, Huntingford C, Wild M, Cox PM (2009) Impact of changes in diffuse radiation on the global land carbon sink. Nature 458:1014–1018. https://doi.org/10.1038/nature07949

Mkhabela MS, Amiro BD, Barr AG, Black TA, Hawthorne I, Kidston J, McCaughey JH, Orchansky AL, Nesic Z, Sass A, Shashkov A, Zha T (2009) Comparison of carbon dynamics and water use efficiency following fire and harvesting in Canadian boreal forests. Agric For Meteorol 149:783–794. https://doi.org/10.1016/j.agrformet.2008.10.025

Myneni RB, Dong J, Tucker CJ, Kaufmann RK, Kauppi PE, Liski J, Zhou L, Alexeyev V, Hughes MK (2001) A large carbon sink in the woody biomass of northern forests. PNAS 98 (26):14784–14789. www.pnas.org/cgi/doi/10.1073/pnas.261555198

Nabuurs G-J, Lindner M, Verkerk PJ, Gunia K, Deda P, Michalak R, Grassi G (2013) First signs of carbon sink saturation in European forest biomass. Nat Clim Chang 3:792–796. https://doi.org/10.1038/NCLIMATE1853

Pan Y, Birdsey RA, Fang J, Houghton R, Kauppi PE, Kurz WA, Phillips OL, Shvindeko A, Lewis SL, Canadell JG, Cias P, Jackson RB, Pacala SW, McGuire AD, Piao S, Rautiainen A, Stich S, Hayes D (2011) A large and persistent carbon sink in the world's forests. Science 333:988–993. https://doi.org/10.1126/science.1201609

Rodda SR, Thumaty KC, Jha CS, Dadhwal VK (2016) Seasonal variations of carbon dioxide, water vapor and energy fluxes in tropical Indian mangroves. Forests 7(35):1–18. https://doi.org/10.3390/f7020035

Rustad LE, Campbell JL, Marion GM, Norby RJ, Mitchell MJ, Hartley AE, Cornelissen JHC, Gurevitch J, GCTE-NEWS (2000) A meta-analysis of the response of soil respiration, net nitrogen mineralization, and aboveground plant growth to experimental ecosystem warming. Oecologia 126:543–562. https://doi.org/10.1007/s004420000544

Saigusa N, Yamamoto S, Hirata R, Ohtani Y, Ide R, Asanuma J, Gamo M, Hirano T, Kondo H, Kosugi Y, Li S-G, Nakai Y, Takagi K, Tani M, Wang H (2008) Temporal and spatial variations in the seasonal patterns of the CO_2 flux in boreal, temperate, and tropical forests in East Asia. Agric For Meteorol 148:700–713

Sakaguchi K, Zeng X, Leung LR, Shao P (2016) Influence of dynamic vegetation on carbon-nitrogen cycle feedback in the community land model (CLM4). Environ Res Lett 11:1–10. https://doi.org/10.1088/1748-9326/aa51d9

Saleska SR, Miller SD, Matross DM, Goulden ML, Wofsy SC, da Racha HR, de Camargo PB, Crill P, Daube BC, de Freitas HC, Hutyra L, Keller M, Kirchhoff V, Menton M, Munger JW, Pyle EH, Rice AH, Silva H (2003) Carbon in Amazon forests: unexpected seasonal fluxes and disturbance-induced losses. Science 302:1554–1557

Schulze E-D, Lloyd J, Kelliher FM, Wirth C, Rebmann C, Lühker B, Mund M, Knohl A, Milyukova M, Schulze W, Ziegler W, Varlagin B, Sogachev F, Valentini R, Dore S, Grigoriev S, Kolle O, Panfyorov MI, Tchebakova N, Vygodskaya NN (1999) Productivity of forests in the Eurosiberian boreal region and their potential to act as carbon sink – a synthesis. Glob Chang Biol 5:703–722

Schulze E-D, Wirth C, Heimann M (2000) Climate change: managing forests after Kyoto. Science 289(5487):2058–2059. www.sciencemag.org/cgi/content/full/sci;289/5487/2058

Smith WK, Reed SC, Cleveland CC, Ballantyne AP, Anderegg WRL, Wieder WR, WR LYY, Running SW (2015) Large divergence of satellite and Earth system model estimates of global terrestrial CO_2 fertilization. Nat Clim Chang 6(3):306–310. https://doi.org/10.1038/NCLIMATE2879

Song X-P, Hansen MC, Stehman SV, Potapov PV, Tyukavina A, Vermote EF, Townshend JR (2018) Global land change from 1982 to 2016. Nature 560:639–643. https://doi.org/10.1038/s41586-018-0411-9

Tan Z, Zhang Y, Yu G, Sha L, Tang J, Deng X, Song Q (2010) Carbon balance of tropical seasonal forest. J Geophys Res 115:1–17

Thornton PE, Doney SC, Lindsay K, Moore JK, Mahowald N, Randerson JT, Fung I, Lamarque J-F, Feddema JJ, Lee Y-H (2009) Carbon-nitrogen interactions regulate climate-carbon cycle feedbacks: results from an atmosphere-ocean general circulation model. Biogeosciences 6:2099–2120

Tian H, Melillo JM, Kicklighter DW, McGuire AD, Helfrich JVK III, Moore B III, Vörösmarty CJ (1998) Effect of interannual climate variability on carbon storage in Amazonian ecosystems. Nature 396:664–667

Valentini R, Matteucci G, Dolman AJ, Schulze E-D, Rebmann C, Moors EJ, Granier A, Gross P, Jensen NO, Pilegaard K, Lindroth A, Grelle A, Bernhofer C, Grünwald T, Aubinet M, Ceulemans R, Kowalski AS, Vesala T, Rannik U, Berbigier P, Loustau D, Guomundsson J, Thorgeirsson H, Ibrom A, Morgenstern K, Clement R, Moncrieff L, Montagnani L, Minerbi S, Jarvis PG (2000) Respiration as the main determinant of carbon balance in European forests. Nature 404:861–865

Vieilledent G, Grinand C, Vaudry R (2013) Forecasting deforestation and carbon emissions in tropical developing countries facing demographic expansion: a case study in Madagascar. Ecol Evol 3(6):1702–1716

Vilén T, Cienciala E, Schelhaas MJ, Verkerk PJ, Lindner M, Peltola H (2015) Increasing carbon sinks in European forests: effects of afforestation and changes in mean growing stock volume. Int J For Res 89:82–90. https://doi.org/10.1093/forestry/cpv034

World Meteorological Organization (2012) The state of greenhouse gases in the atmosphere based on global observations through 2011. WMO Greenhouse Gas Bulletin 8:1–4. ISSN 2078-0796. https://www.wmo.int/pages/prog/arep/gaw/ghg/ghg9-en-online.html

World Meteorological Organization (2013) The state of greenhouse gases in the atmosphere using global observations through 2012. WMO Greenhouse Gas Bulletin 9. https://www.wmo.int/pages/prog/arep/gaw/ghg/ghg9-en-online.html

World Meteorological Organization (2014) The state of greenhouse gases in the atmosphere using global observations through 2013. WMO Greenhouse Gas Bulletin 10:1–8. ISSN 2078-0796

Xiao J, Liu S, Stoy PC (2016) Preface: impacts of extreme climate events and disturbances on carbon dynamics. Biogeosciences 13:3665–3675. https://doi.org/10.5194/bg-13-3665-2016

Zhao M, Running SW (2010) Drought-induced reduction in global terrestrial net primary production from 2000 through 2009. Science 329:940–943. https://doi.org/10.1126/science.1192666

Chapter 5
Carbon Sequestration via Biomineralization: Processes, Applications and Future Directions

Willis Gwenzi

Abstract Carbon dioxide (CO_2) emissions from anthropogenic and natural sources contribute to greenhouse gas emissions causing climate change and variability. In recent years, microbially-induced biomineralization of CO_2 to precipitate calcium carbonate has been widely explored as a potential strategy to sequester carbon. Several microorganisms occurring in marine, lacustrine and terrestrial ecosystems, including cyanobacteria have capacity to sequester carbon dioxide via biomineralization. Similarly, a number of plant species sequester carbon dioxide through biomineralization. The objectives of this chapter are: (1) to summarize the processes and mechanisms of microbially- and plant-mediated biomineralization, (2) highlight potential applications of biomineralization in CO_2 sequestration, and (3) highlight key knowledge gaps and future research directions.

Microbially-mediated calcium carbonate precipitation occurs via three main mechanisms: (1) microbially-controlled, (2) microbially-induced, and (3) biologically-influenced biomineralization. Several metabolic processes control biomineralization, including: (1) urea hydrolysis, (2) activity of carbonic anhydrase, (3) microbial reduction of Fe(III), (4) photosynthesis by cyanobacteria fixes CO_2, and (5) sulphate reduction. Microbial mats and extracellular polymeric substances (EPS) also play a critical role in microbially-mediated biomineralization. Plant species such as the oxalogenic iroko tree (*Milicia excelsa*) common in coastal tropical Africa, and several acacia species in Australia have been reported to sequester carbon dioxide via the oxalate-carbonate pathway. Calcium carbonate has a longer residence time in the order of millennia, thus is a more stable carbon pool than organic carbon in soils and live biomass. Accordingly, calcium carbonate formed via biomineralization serves a dual function: (1) in carbon capture and storage systems, microbially mediated biomineralization is used as a benign sealant to avoid escape of CO_2 directly injected into deep geological systems, and (2) both microbially-and plant-mediated biomineralization sequester CO_2, thus reducing its concentration in the atmosphere. For example, an 80 year-old iroko tree is estimated to sequester 500 kg of carbon in its trunk, and approximately 1000 kg of carbon in

W. Gwenzi (✉)
Biosystems and Environmental Engineering Research Group, Department of Soil Science and Agricultural Engineering, Faculty of Agriculture, University of Zimbabwe, Harare, Zimbabwe

Inamuddin et al. (eds.), *Sustainable Agriculture Reviews 37*, Sustainable Agriculture Reviews 37, https://doi.org/10.1007/978-3-030-29298-0_5

the surrounding soil, giving a total of 1500 kg of carbon. Microbially-mediated biomineralization is also used in construction materials, cementation and stabilization of porous materials, sub-surface barriers, aquacultural ponds, industrial filler materials, hydraulic control and environmental remediation. Although not designed to store carbon, these applications also constitute a stable carbon pool, thus contribute to carbon sequestration. However, large scale commercialization of the technology remain low, while the economics, and long-term behaviour of biominerals remain poorly understood.

Keywords Biomineralization · Calcium carbonate precipitation · Carbon capture and storage · Anthropogenic greenhouse gases · Microbially mediated biomineralization · Occluded carbon · Organomineralization · Plant-mediated biomineralization · Sequestration · Stable carbon pools

Acronyms

Ca	calcium
Ca^{2+}	calcium ions
$CaCO_3$	calcium carbonate
CCS	carbon capture and storage
CO_2	carbon dioxide
EPS	Extracellular polymeric substances
Fe(III)	iron (Fe) in oxidation state of three
$FeCO_3$	Ferrous (Fe(II)) carbonate also known as siderite
GT	gigatonnes equivalent to 10^9 tonnes
HCO_3^-	bicarbonate or hydrogen carbonate ion
CO_3^{2-}	carbonate ion
Mg	Magnesium
Mt.	million (10^6) tonnes
OH^-	Hydroxyl ion
PCB	polychlorinated biphenyls

5.1 Introduction

Carbon dioxide emissions from anthropogenic and natural sources contribute to greenhouse gases causing climate change and variability. Anthropogenic sources of carbon dioxide emissions include; combustion of fossil and biomass fuels, industrial processes such as cement production, deforestation and agro-ecosystems including crop and livestock production systems. For example, anthropogenic sources, including combustion of fossil fuels and cement production contribute approximately 9 gigatonnes (GT) of carbon per year (equivalent to 33 GT CO_2). Considering that terrestrial and oceanic systems absorb approximately 5 GT of carbon per year, approximately 4 GT of carbon remain in the atmosphere (Jansson et al. 2010). Climate change and variability is associated with rising global atmospheric temperature, changes in rainfall patterns and intensity, and diseases and pest outbreaks.

Recent years have witnessed a concerted effort to develop and adapt mitigation interventions to reduce anthropogenic emissions, and increase carbon capture and storage. These mitigation measures include: (1) carbon capture and storage (CCS), entailing carbon capture from point sources such as thermal power plants, and subsequent direct injection of CO_2 into geological formations (Boot-Handford et al. 2014; Rubin and De Coninck 2005), (2) reducing deforestation, and promoting soil carbon sequestration and organic carbon sequestration in plants via photosynthesis (Kindermann et al. 2008; Lal 2004), and (3) a shift from non-renewable fossil fuels such as coal and petroleum-based fuels to renewable energy sources, including biofuels and solar energy (Sadorsky 2009; Menyah and Wolde-Rufael 2010). The effects of these mitigation measures on anthropogenic emissions have been the subject several reviews (Snyder et al. 2009; Lenzen 2008; Von Blottnitz and Curran 2007).

In order to have significant impact on greenhouse gas mitigation, carbon dioxide (CO_2) must be captured and sequestered over very long (geologic) time intervals. Therefore, carbon sequestration in soils and live plant biomass can be considered as short-term or unstable pools of carbon, because such organic carbon often has a limited residence time. Accordingly, recent attention on mitigation of greenhouse gas emission has shifted towards direct injection of CO_2 from point sources into deep geological formations (Dhami et al. 2013). However, ensuring long-term storage of CO_2 in such geological systems could be problematic due to leakages. The escape of large quantities of CO_2 from such geological systems into the atmosphere could have devastating effects on climate (Benson and Orr 2008). To address these limitations, microbially-mediated biomineralization or organomineralization has been proposed as a possible long-term solution to sequester carbon dioxide (Cailleau et al. 2004; Dupraz et al. 2004, 2009a). Biomineralization refers to the process whereby organisms form biominerals through biologically-induced or biologically-controlled processes (Fig. 5.1) (Dupraz et al. 2009a; He et al. 2014).

Biomineralization occurs in a wide range of environments, including marine, lacustrine and terrestrial ecosystems. To date, microbially-mediated biomineralization has been widely reported in several microorganisms, including; cyanobacteria, methanogenic archaea, anoxygenic phototrophs, sulfate-reducing bacteria, and heterotrophic ureolytic bacteria (Bundeleva et al. 2012). Moreover, studies documenting biomineralization in higher plants dates back to the 1930s, when calcium carbonate crystals or 'stones' were reported in an oxalogenic iroko tree (Campbell and Fisher 1932; Harris 1933). Later studies have estimated the contribution of the iroko tree to carbon dioxide sequestration, and the mechanism involved (Verrecchia et al. 2006; Cailleau et al. 2011; Aragno and Verrecchia 2012). Moreover, biomineralization has been reported in several other plant species including acacia species and field crops (Ruiz et al. 2002; Mazen 2004; Mazen et al. 2004; Nitta et al. 2006; He et al. 2012a, b, 2014). Although carbon dioxide

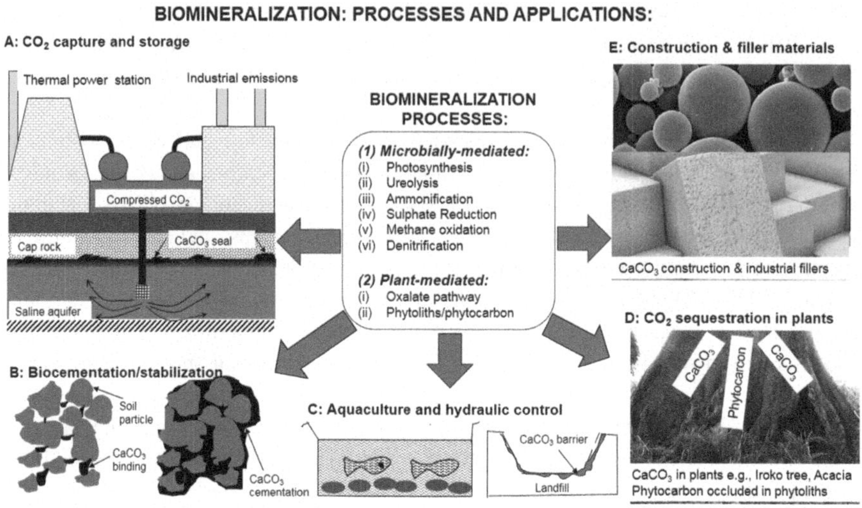

Fig. 5.1 Conceptual depiction of the microbially- and plant-mediated biomineralization processes, and potential applications of biomineralization

biomineralization appears common in several microbial and plant species, its contribution to carbon sequestration is currently under-estimated (Cailleau et al. 2011).

The current chapter seeks to highlight the potential of carbon sequestration via two categories of biomineralization; (1) microbially-mediated biomineralization, and (2) plant-mediated biomineralization. The objectives are: (1) to summarize the biogeochemical processes and mechanisms involved in microbially- and plant-mediated biomineralization in terrestrial ecosystems; (2) highlight the nature, properties and behaviour and fate of biominerals; (3) discuss the current and potential applications of biomineralization in bioenvironmental engineering, and (4) identify key knowledge gaps and future research directions.

5.2 Biomineralization Processes and Mechanisms

5.2.1 Microbially-Mediated Biomineralization

Microbially-mediated carbon dioxide biomineralization occurs in several microbial species found in marine, lacustrine and terrestrial ecosystems, including cyanobacteria (Dupraz et al. 2009a, b, c). The formation of metal carbonates via ureolytic biomineralization has also been reported in urease-positive fungus *Neurospora crassa*, and a detailed discussion of the mechanism is presented in literature (Li et al. 2014). Typical carbon-rich biominerals include; calcium carbonate, magnesium carbonate, calcium phosphate carbonate and siderite ($FeCO_3$) (Rivadeneyra et al. 2010; Roh et al. 2003). Three anhydrous polymorphs of calcium carbonate (calcite, aragonite and vaterite) and two hydrated crystalline phases (monohydrocalcite ($CaCO_3 \cdot H_2O$) and ikaite ($CaCO_3 \cdot 6H_2O$)) are formed through microbially-mediated biomineralization (Dhami et al. 2013). The microbial processes and biogeochemical factors controlling biomineralization has been discussed in several reviews (Dupraz et al. 2004, 2009a; Benzerara et al. 2011; Jansson et al. 2010).

The three dominant mechanisms of microbial biomineralization include (Dupraz et al. 2009a):

1. Microbially-controlled biomineralization, whereby specific cellular activities control the nucleation, growth, morphology, and final location of a biomineral;
2. Microbially-induced biomineralization arising from the indirect modification of physico-chemical conditions. In microbially-induced processes, biominerals are formed as a by-product of microbial metabolic processes such as urea hydrolysis (ureolytic biomineralization), sulphate reduction, nitrate reduction and anaerobic activities that affect pH, and concentrations and super-saturation of carbonates;
3. Biologically-influenced biomineralization, which is the passive mineral precipitation occurring in the presence of organic materials. These organic materials include extracellular polymeric substances (EPS) or cell surfaces, whose properties determine crystal composition and morphology.

Specific metabolic processes that control microbially-induced biomineralization are summarised in Fig. 5.1. These processes include, among others (Dhami et al. 2013; Dupraz and Visscher 2005):

1. Urea hydrolysis or ureolytic biomineralization, whereby urea hydrolysis generates hydroxyl (OH^-) ions that shift equilibrium towards the formation of bicarbonate (HCO_3^-) and carbonate (CO_3^{2-}) ions. In the presence of significant amounts of calcium ions (Ca^{2+}) calcium carbonate ($CaCO_3$) precipitation occurs (Phillips et al. 2013).
2. Carbonic anhydrase biomineralization involving enzyme-catalysed hydration of carbon dioxide (CO_2) to carbonates (Tambutté et al. 2007; Kim et al. 2012).
3. Microbial reduction of ferric ions (Fe(III)), resulting in the simultaneous formation of siderite ($FeCO_3$) and magnetite (Fe_3O_4)
4. Photosynthesis by cyanobacteria fixes CO_2, thus high rates of photosynthesis deplete carbon dioxide (CO_2) and increase alkalinity which in turn promotes the precipitation of $CaCO_3$.
5. Sulphate reduction promotes the precipitation of calcium carbonate.

In summary, cyanobacterial photosynthesis and sulphate reduction result in calcium carbonate ($CaCO_3$) precipitation, while aerobic respiration, fermentation and sulphate oxidation promote dissolution (Dhami et al. 2013). Considering these counteracting processes, calcium carbonate formation only occurs when processes promoting precipitation exceed dissolution. Several physico-chemical properties, including pH and biogeochemical processes also control biomineralization. For example, studies also show that microbial mats in marine and lacustrine ecosystems are biogeochemical hotspots, exhibiting high metabolic rates, including photosynthesis, aerobic respiration and sulphate reduction (Dupraz and Visscher 2005).

Microbial mats are a type of biofilms consisting of vertically laminated sedimentary material occurring in various environments, including lagoons, marine, hypersaline ponds, hot springs and fresh water aquatic systems (Visscher et al. 1992). The functional groups of microbes commonly found in microbial mats, and their role in biomineralization are described in literature (van Gemerden 1993; Dupraz and Visscher 2005).

5.2.2 Plant-Mediated Biomineralization

Several plant species concentrate biominerals, including oxalate salts, calcium carbonates and silicates (Mazen 2004; Mazen et al. 2004; He et al. 2014). An example is the oxalogenic iroko tree (*Milicia excelsa*) native to coastal tropical African countries, including Angola, Benin, Cameroon, Congo, Ghana, Guinea and Mozambique. The occurrence of biominerals in the iroko tree has a long history dating back to the 1930s, when calcium carbonate crystals were observed in iroko wood (Harris 1933). Recently, studies conducted in Cameroun and Ivory Coast elucidated the mechanisms of biomineralization and the potential contribution of the iroko tree to carbon sequestration (Braissant et al. 2004; Verrecchia et al. 2006; Aragno and Verrecchia 2012). In the same studies, it has been shown that biomineralization in the iroko tree occurs via the oxalate-carbonate pathway. Cailleau et al. (2011) present a detailed discussion of the oxalate-carbonate pathway in the iroko tree. Biomineralization also occurs in several other plant species including acacia species, and even field crops (Mazen 2004; Mazen et al. 2004; Nitta et al. 2006; He et al. 2014). The physiological basis for biomineralization has been discussed in detail in an earlier review paper (He et al. 2014). In summary, the ecophysiological significance of biomineralization include (He et al. 2014): (1) regulating the cytoplasmic concentration of free calcium, (2) inactivating phytotoxic toxic elements such as aluminium and other metals, (3) light acquisition and scattering to optimize photosynthesis, (4) aiding in pollen release, germination, and tube growth, (5) as a deterrent to herbivory, and (6) as a mechanism of biogeochemical cycling of carbon, calcium, and silicon, including sequestration of atmospheric carbon dioxide (CO_2).

5.3 Carbon Dioxide Sequestration

5.3.1 Microbially-Mediated Biomineralization

To this point, the chapter has demonstrated that several microbially-mediated processes contribute to carbon biomineralization (Fig. 5.1). The carbon-rich minerals have residence times in the order of geological timescales, and represent a stable pool of carbon. For example, there is supporting evidence that massive carbonate mineral reservoirs similar to that formed via microbially-mediated biomineralization have existed for long geological timescales spanning millions of years (Dhami et al.

2013). This is in contrast to organic carbon in living organisms, which often has a low residence time. Therefore, capturing and sequestering the carbon dioxide (CO_2) via biomineralization is often considered a safe and more practical way for reducing greenhouse gas emissions (Shaffer 2010; Sharma and Bhattacharya 2010). This method entails fixing the CO_2 in the stable pools of carbonate minerals such as calcite, magnesite, and dolomite in deep geologic reservoirs, such as saline aquifers. Although large quantities of calcium carbonate exist in marine, lacustrine and terrestrial systems, the exact total quantity of carbon dioxide (CO_2) sequestered via microbially-mediated biomineralization is not known.

Carbon dioxide (CO_2) directly injected into geological systems is prone to release back into the atmosphere through leakages induced by poor integrity of geological systems, the corrosive effect or carbonation of supercritical CO_2, and increased permeability of the capping rock (Barlet-Gouédard et al. 2009; Wigand et al. 2009; Carey et al. 2010). Conventional sealing materials such as cements often have higher viscosity than the surrounding aqueous solution, thus fail to effectively seal micropore spaces and microfractures, through which low viscosity supercritical CO_2 could leak. In this regard, biomineralization can be used as an effective sealing method to prevent CO_2 escape from such geological systems (Fig. 5.1). Three methods based on ureolytic biomineralization have been suggested to reduce CO_2 leakages: (1) formation trapping, (2) solubility trapping, and (3) mineral trapping (Dupraz et al. 2009b, c; Mitchell et al. 2010). In natural systems, the fixation of CO_2 in the form of calcite, aragonite, dolomite and magnesite occurs over geological time scales. Therefore, to hasten the biomineralization process, researchers have proposed the use of biomimetic carbon dioxide (CO_2) sequestration, where biological catalysts or enzymes such as carbonic anhydrase are used to reduce the localized concentration of CO_2 concentration (Smith and Ferry 2000). Carbonic anhydrase is ubiquitous in several organisms and critical to many biological processes, including photosynthesis, respiration, and CO_2 and ion transport (Smith and Ferry 2000; Liu et al. 2005). In fact, the role of carbonic anhydrase in biomineralization in marine, lacustrine and terrestrial organisms is well-known although its application for carbon dioxide sequestration is fairly recent (Beier and Anken 2006; Tambutté et al. 2007). Sequestration of CO_2 by carbonic anhydrase isolated from various bacterial species has been reported in several studies (Ramanan et al. 2009; Yadav et al. 2011; Dhami et al. 2013). For example, Liu et al. (2005) used carbonic anhydrase derived from bovines to accelerate CO_2 hydration, and observed that the precipitation of calcium carbonate was much faster in the presence of the bovine enzyme than without. Kim et al. (2012) used a recombinant carbonic anhydrase to enhance calcium carbonate precipitation. Mitchell et al. (2010) and Phillips et al. (2013) applied calcium carbonate biomineralization to protect the well cements from supercritical CO_2 by plugging microfractures near wells, while reducing the permeability of the capping rock. In other studies, *S. pasteurii* was used to determine the transformation of CO_2 into a solid carbonate phase or mineral trapping in artificial groundwater (Dupraz et al. 2009b, c). In another study, Jansson and Northen (2010) applied cyanobacteria for carbon capture and sequestration from point sources, while Kupriyanova et al. (2007) successful deposited $CaCO_3$ by extracellular carbonic anhydrase derived from cyanobacterial cells. The cyanobacteria use solar energy during photosynthesis to transform carbon dioxide to recalcitrant calcium carbonate (Kamennaya et al. 2012). These studies demonstrate that, indeed microbially mediated biomineralization has the potential to capture and sequester carbon in stable inorganic carbon pools. In summary, calcium carbonate plays a dual role; first, as a stable pool of carbon, and second, as an effective sealant to prevent CO_2 release back into the atmosphere.

Besides direct carbon sequestration, other biomineralization applications that indirectly sequester carbon are presented in literature (Phillips et al. 2013). Figure 5.1 presents a summary of the various applications. For example, biomineralization can be used for biodeposition and biocement in construction applications (Dhami et al. 2013). In addition, biomineralization is used to stabilize porous materials, including sands and liquefiable soils, and in sub-surface barriers and aquacultural ponds. Moreover, biomineralization is used in hydraulic control and environmental remediation of soils and sediments contaminated with radionuclides, metals and hydrocarbons such as polychlorinated biphenyls (PCBs) (Phillips et al. 2013). Microbially-mediated calcium carbonate precipitation is also used to develop ideal materials that can be used as fillers in industrial materials such as rubber and plastics, fluorescent particles in stationery inks and fluorescent markers (Yoshida et al. 2010; Ghami et al. 2013). These materials, though not intentionally designed to store carbon, represent a potential reservoir of stable carbon in the form of calcium carbonates.

5.3.2 *Plant-Mediated Biomineralization*

5.3.2.1 The Case of the Iroko Tree

The iroko tree in West Africa (Cameroun and Ivory Coast) represents one case where biomineralization was investigated in the context of carbon sequestration and mitigation of greenhouse emissions. Braissant et al. (2004) estimated that an 80-year old iroko tree sequestered approximately 500 kg of inorganic carbon as calcium carbonate in its trunk, and a corresponding 1000 kg of carbon in the surrounding soil. Taken together, on average, an 80-year iroko tree sequesters a total of approximately 1500 kg of carbon, equivalent to 20 kg of carbon/year. However, Braissant and co-workers did not include calcium carbonate stored in the leaves and roots of the same tree, thus could have under-estimated the carbon sequestration of the iroko tree. Moreover, the carbon sequestration potential of the iroko tree could be even higher given that such trees could live for up to 500 years. Although data only exist for the iroko tree, carbon oxalate biosynthesis occurs in a wide range of plant and even fungal species (Li et al. 2014). Therefore, biomineralization via the oxalate-carbon pathway could be a more prevalent phenomenon than initially thought.

5.3.2.2 The Case of Australian Acacia Species

In the Great Sandy Desert in north-western Australia, He et al. (2012a, b, 2014) investigated the morphology and precipitation of biominerals in phyllodes and branchlets of four acacia species: (1) *Acacia robeorum*, (2) *Acacia ancistrocarpa*, (3) *Acacia stipuligera*, and (4) *Acacia stellaticeps*. In the same studies, the authors reported that the most common carbon biominerals were carbon oxalate and calcium carbonate. However, the work of He and co-workers was limited to the ecophysiological significance of biominerals, thus provides no estimates of total carbon sequestration at plant, plot or ecosystem scales. Moreover, although He et al. (2012a, b) investigated just four species, there is a possibility that biomineralization may also occur in other co-occurring species. For example, biomineralization appears prevalent in other acacia species in arid and semi-arid environments, and has been reported in other studies conducted outside Australia (Brown et al. 2013). Barring acacia species, and arid and semi-arid environments, biominerals, including carbon oxalates and calcium carbonates have been documented in duckweed, sugar beet (*Beta vulgaris*), silky glycine (*Glycine canescens*), pink fringe (*Arthrostema ciliatum*) (Franceschi and Schueren 1986) and water hyacinth (*Eichhornia crassipes*) (Mazen and El Maghraby 1998). A review by He et al. (2014) presents a comprehensive list of other plant species where biomineralization has been reported. In summary, these studies point to the potential of carbon biomineralization being a common phenomenon in plant species. However, further studies are required to understand whether the phenomenon is also common in most plants, and estimate its contribution to carbon sequestration at plant, plot and ecosystem scales so that its contribution to the global carbon cycle is accorded its rightful place. The consideration of carbon biomineralization is important given that calcium carbonate has a higher residence time than organic carbon in soils and live biomass. In this regard, biominerals may represent a stable carbon pool especially in tropical, semi-arid and arid environments where soil organic matter turnover may be high due to high oxidation rates associated with high temperatures and strong wetting and drying cycles.

5.3.2.3 Carbon Occlusion in Biominerals

Besides the formation of calcium carbonate and other carbonate minerals such as calcium phosphate carbonate, biomineralization may also result in the formation of silicate mineral or silica. Such silica minerals often form structures in plants called phytolith (Parr and Sullivan 2005, 2011). Phytoliths can occlude carbon and thus stabilize it. In India, one study estimated that phytoliths in agronomic crops occluded 87 million tonnes (Mt) of phytocarbon per year, thus contribute to carbon sequestration (Rajendiran et al. 2012). Like biomineralization, carbon occlusion in phytoliths also occurs in several plant species, including field crops such as sugarcane and rice (Parr and Sullivan 2005; Parr et al. 2009; Rajendiran et al. 2012, 2016). However, such carbon sequestration mechanisms are often excluded in conventional carbon accounting for agro-ecosystems. Therefore, there is need to further estimate the contribution of phytocarbon to carbon sequestration at various spatio-temporal scales. Understanding the contribution of phytocarbon to carbon sequestrations at species, plant, plot, farm, country and global scales may provide critical information on the choice of agro-ecosystems to optimize carbon sequestration. However, there is need to quantify the residence times of such phytocarbon. In this regards, isotopic tracing studies may provide critical information on the partitioning and stability of phytocarbon.

5.4 Knowledge Gaps and Future Directions

The chapter highlighted that biomineralization is common in microbial and plant species, and occurs in diverse environments, including marine, lacustrine and terrestrial ecosystems. However, future research is required to address the following knowledge gaps:

1. Large-scale applications are yet to be undertaken, thus data on its economic viability is still lacking;
2. The microbial and plant species reported in literature may represent a small fraction of the total species with capacity for biomineralization, thus additional research is required to identify other species with similar capacity;
3. The oxalate-carbonate pathway appears to be the dominant carbon biomineralization mechanism in plants such as the iroko tree and acacia species, but it remains unclear whether other mechanisms also contribute.
4. The genetic basis of biomineralization is still poorly understood, hence there is need to investigate the potential to develop microbial and plant species with high carbon biomineralization capacity. Recent advances in genomic and metagenomic techniques may provide further insights in this regard.
5. Carbon mineralization in plant species is limited to individual plants, and in some cases mere detection of its presence, while studies investigating such phenomenon at ecosystem and global scales are still limited. This raises the question whether biomineralization observed at individual plant scales can be reproduced at large or human-relevant scales.
6. Limited data exist on the longevity and biogeochemical behaviour of biominerals, and their fate in the environment. Tracer studies using isotopes and rare earth elements may provide some insights on the behaviour and fate of biominerals.

5.5 Summary and Conclusions

The current chapter discussed the mechanisms of microbially- and plant-mediated biomineralization, and its potential application for carbon dioxide sequestration. Several microorganisms occurring in marine, lacustrine and terrestrial ecosystems, including cyanobacteria sequester carbon dioxide via biomineralization. Microbially-mediated calcium carbonate precipitation occurs via three main mechanisms: (1) microbially-controlled, (2) microbially-induced, and (3) biologically-influenced biomineralization.

Several metabolic processes control biomineralization, including; urea hydrolysis, activity of carbonic anhydrase, microbial reduction of ferric iron (Fe(III)), photosynthesis by cyanobacteria, and sulphate reduction. Microbial mats and extracellular polymeric substances (EPS) also play a critical role in microbially-mediated biomineralization, especially in cyanobacteria. Biomineralization via the oxalate-carbonate pathway is also prevalent among plant species, including the oxalogenic iroko tree (*Milicia excelsa*), several acacia species in Australia and even agronomic crops. Due to its long residence time in the order of millennia, calcium carbonate represents a more stable carbon pool than organic carbon in soils and live biomass. Accordingly, calcium carbonate directly contributes to carbon sequestration, and is also used to prevent the escape of carbon dioxide (CO_2) directly injected into deep geological systems. Moreover, the application of biomineralization in construction materials, cementation and stabilization of porous materials, sub-surface barriers, aquacultural ponds, industrial filler materials, hydraulic control and environmental remediation constitutes an indirect method to carbon sequestration. However, further research, including economic analysis is required to address several knowledge gaps before large-scale commercialization of the technology. Addressing these knowledge gaps presents an advancement in the biomineralization technology, which could result in its large scale application for carbon dioxide (CO_2) sequestration.

References

Aragno M, Verrecchia E (2012) The oxalate-carbonate pathway: a reliable sink for atmospheric CO_2 through calcium carbonate biomineralization in ferralitic tropical soils. In: Microorganisms in environmental management. Springer, Dordrecht, pp 191–199. https://doi.org/10.1007/978-94-007-2229-39

Barlet-Gouédard V, Rimmelé G, Porcherie O, Quisel N, Desroches J (2009) A solution against well cement degradation under CO_2 geological storage environment. Int J Greenh Gas Con 3 (2):206–216. https://doi.org/10.1016/j.ijggc.2008.07.005

Beier M, Anken R (2006) On the role of carbonic anhydrase in the early phase of fish otolith mineralization. Adv Space Res 38(6):1119–1122. https://doi.org/10.1016/j.asr.2005.10.027

Benson SM, Orr FM (2008) Carbon dioxide capture and storage. MRS Bull 33(4):303–305. https://doi.org/10.1557/mrs2008.63

Benzerara K, Miot J, Morin G, Ona-Nguema G, Skouri-Panet F, Ferard C (2011) Significance, mechanisms and environmental implications of microbial biomineralization. C R Geosci 343 (2–3):160–167. https://doi.org/10.1016/j.crte.2010.09.002

Boot-Handford ME, Abanades JC, Anthony EJ, Blunt MJ, Brandani S, Mac Dowell N, Fernández JR, Ferrari MC, Gross R, Hallett JP, Haszeldine RS (2014) Carbon capture and storage update. Energ Environ Sci 7(1):130–189. https://doi.org/10.1039/C3EE42350F

Braissant O, Cailleau G, Aragno M, Verrecchia EP (2004) Biologically induced mineralization in the tree *Milicia excelsa* (Moraceae): its causes and consequences to the environment. Geobiology 2(1):59–66. https://doi.org/10.1111/j.1472-4677.2004.00019.x

Brown SL, Warwick NW, Prychid CJ (2013) Does aridity influence the morphology, distribution and accumulation of calcium oxalate crystals in Acacia (Leguminosae: Mimosoideae)? Plant Physiol Biochem 73:219–228. https://doi.org/10.1016/j.plaphy.2013.10.006

Bundeleva IA, Shirokova LS, Bénézeth P, Pokrovsky OS, Kompantseva EI, Balor S (2012) Calcium carbonate precipitation by anoxygenic phototrophic bacteria. Chem Geol 291:116–131. https://doi.org/10.1016/j.chemgeo.2011.10.003

Cailleau G, Braissant O, Verrecchia EP (2004) Biomineralization in plants as a long-term carbon sink. Naturwissenschaften 91(4):191–194. https://doi.org/10.1007/s00114-004-0512-1

Cailleau G, Braissant O, Verrecchia EP (2011) Turning sunlight into stone: the oxalate-carbonate pathway in a tropical tree ecosystem. Biogeosci Discuss 8(1):1077–1108. https://doi.org/10.5194/bg-8-1755-2011

Campbell WG, Fisher RC (1932) The composition and origin of "stone" in the iroko wood. Empir For J 11(2):244–245. https://www.jstor.org/stable/42595822

Carey JW, Svec R, Grigg R, Zhang J, Crow W (2010) Experimental investigation of wellbore integrity and CO_2–brine flow along the casing–cement microannulus. Int J Greenh Gas Con 4 (2):272–282. https://doi.org/10.1016/j.ijggc.2009.09.018

Dhami NK, Reddy MS, Mukherjee K (2013) Biomineralization of calcium carbonates and their engineered applications: a review. Front Microbiol 4:314. https://doi.org/10.3389/fmicb.2013.00314

Dupraz C, Visscher PT (2005) Microbial lithification in marine stromatolites and hypersaline mats. Trends Microbiol 13(9):429–438. https://doi.org/10.1016/j.tim.2005.07.008

Dupraz C, Visscher PT, Baumgartner LK, Reid RP (2004) Microbe–mineral interactions: early carbonate precipitation in a hypersaline lake (Eleuthera Island, Bahamas). Sedimentology 51 (4):745–765. https://doi.org/10.1111/j.1365-3091.2004.00649.x

Dupraz C, Reid RP, Braissant O, Decho AW, Norman RS, Visscher PT (2009a) Processes of carbonate precipitation in modern microbial mats. Earth Sci Rev 96(3):141–162. https://doi.org/10.1016/j.earscirev.2008.10.005

Dupraz S, Ménez B, Gouze P, Leprovost R, Bénézeth P, Pokrovsky OS, Guyot F (2009b) Experimental approach of CO_2 biomineralization in deep saline aquifers. Chem Geol 265 (1–2):54–62. https://doi.org/10.1016/j.chemgeo.2008.12.012

Dupraz S, Parmentier M, Ménez B, Guyot F (2009c) Experimental and numerical modeling of bacterially induced pH increase and calcite precipitation in saline aquifers. Chem Geol 265 (1–2):44–53. https://doi.org/10.1016/j.chemgeo.2009.05.003

Franceschi VR, Schueren AM (1986) Incorporation of strontium into plant calcium oxalate crystals. Protoplasma 130:199–205. https://doi.org/10.1007/BF01276601

Harris CM (1933) Stone in *Chlorophora excelsa* B. & H. iroko. Empir For J 12:229–238. https://www.jstor.org/stable/42595874

He H, Bleby TM, Veneklaas EJ, Lambers H, Kuo J (2012a) Precipitation of calcium, magnesium, strontium and barium in tissues of four Acacia species (Leguminosae: Mimosoideae). PLoS One 7(7):e41563. https://doi.org/10.1371/journal.pone.0041563

He H, Bleby TM, Veneklaas EJ, Lambers H, Kuo J (2012b) Morphologies and elemental compositions of calcium crystals in phyllodes and branchlets of *Acacia robeorum* (Leguminosae: Mimosoideae). Ann Bot 109(5):887–896. https://doi.org/10.1093/aob/mcs004

He H, Veneklaas EJ, Kuo J, Lambers H (2014) Physiological and ecological significance of biomineralization in plants. Trends Plant Sci 19(3):166–174. https://doi.org/10.1016/j.tplants.2013.11.002

Jansson C, Northen T (2010) Calcifying cyanobacteria—the potential of biomineralization for carbon capture and storage. Curr Opin Biotechnol 21(3):365–371. https://doi.org/10.1016/j.copbio.2010.03.017

Jansson C, Wullschleger SD, Kalluri UC, Tuskan GA (2010) Phytosequestration: carbon biosequestration by plants and the prospects of genetic engineering. Bioscience 60 (9):685–696. https://doi.org/10.1525/bio.2010.60.9.6

Kamennaya N, Ajo-Franklin C, Northen T, Jansson C (2012) Cyanobacteria as biocatalysts for carbonate mineralization. Fortschr Mineral 2(4):338–364. https://doi.org/10.3390/min2040338

Kim IG, Jo BH, Kang DG, Kim CS, Choi YS, Cha HJ (2012) Biomineralization-based conversion of carbon dioxide to calcium carbonate using recombinant carbonic anhydrase. Chemosphere 87 (10):1091–1096. https://doi.org/10.1016/j.chemosphere.2012.02.003

Kindermann G, Obersteiner M, Sohngen B, Sathaye J, Andrasko K, Rametsteiner E, Schlamadinger B, Wunder S, Beach R (2008) Global cost estimates of reducing carbon emissions through avoided deforestation. Proc Natl Acad Sci 105(30):10302–10307. https://doi.org/10.1073/pnas.0710616105

Kupriyanova E, Villarejo A, Markelova A, Gerasimenko L, Zavarzin G, Samuelsson G, Los DA, Pronina N (2007) Extracellular carbonic anhydrases of the stromatolite-forming cyanobacterium Microcoleus chthonoplastes. Microbiology 153(4):1149–1156. https://doi.org/10.1099/mic.0.2006/003905-0

Lal R (2004) Soil carbon sequestration impacts on global climate change and food security. Science 304(5677):1623–1627. https://doi.org/10.1126/science.1097396

Lenzen M (2008) Life cycle energy and greenhouse gas emissions of nuclear energy: a review. Energ Conver Manage 49(8):2178–2199. https://doi.org/10.1016/j.enconman.2008.01.033

Li Q, Csetenyi L, Gadd GM (2014) Biomineralization of metal carbonates by *Neurospora crassa*. Environ Sci Technol 48(24):14409–14416. https://doi.org/10.1021/es5042546

Liu N, Bond GM, Abel A, McPherson BJ, Stringer J (2005) Biomimetic sequestration of CO_2 in carbonate form: role of produced waters and other brines. Fuel Process Technol 86 (14–15):1615–1625. https://doi.org/10.1016/j.fuproc.2005.01.008

Mazen AMA (2004) Calcium oxalate deposits in leaves of *Corchorus olitorius* as related to accumulation of toxic metals. Russ J Plant Physiol 51:281–285. https://doi.org/10.1023/B:RUPP.0000019226.03536.21

Mazen AMA, El Maghraby OMO (1998) Accumulation of cadmium, lead and strontium, and a role of calcium oxalate in water hyacinth tolerance. Biol Plant 40:411–417. https://doi.org/10.1023/A:1001174132428

Mazen AM, Zhang D, Franceschi VR (2004) Calcium oxalate formation in Lemna minor: physiological and ultrastructural aspects of high capacity calcium sequestration. New Phytol 161 (2):435–448. https://doi.org/10.1046/j.1469-8137.2003.00923.x

Menyah K, Wolde-Rufael Y (2010) CO_2 emissions, nuclear energy, renewable energy and economic growth in the US. Energy Policy 38(6):2911–2915. https://doi.org/10.1016/j.enpol.2010.01.024

Mitchell AC, Dideriksen K, Spangler LH, Cunningham AB, Gerlach R (2010) Microbially enhanced carbon capture and storage by mineral-trapping and solubility-trapping. Environ Sci Technol 44(13):5270–5276. https://doi.org/10.1021/es903270w

Nitta I, Kida A, Fujibayashi Y, Katayama H, Sugimura Y (2006) Calcium carbonate deposition in a cell wall sac formed in mulberry idioblasts. Protoplasma 228(4):201–208. https://doi.org/10.1007/s00709-006-0182-2

Parr JF, Sullivan LA (2005) Soil carbon sequestration in phytoliths. Soil Biol Biochem 37 (1):117–124. https://doi.org/10.1016/j.soilbio.2004.06.013

Parr JF, Sullivan LA (2011) Phytolith occluded carbon and silica variability in wheat cultivars. Plant and Soil 342(1–2):165–171. https://doi.org/10.1007/s11104-010-0680-z

Parr J, Sullivan L, Quirk R (2009) Sugarcane phytoliths: encapsulation and sequestration of a long-lived carbon fraction. Sugar Tech 11(1):17–21. https://doi.org/10.1007/s12355-009-0003-y

Phillips AJ, Gerlach R, Lauchnor E, Mitchell AC, Cunningham AB, Spangler L (2013) Engineered applications of ureolytic biomineralization: a review. Biofouling 29(6):715–733. https://doi.org/10.1080/08927014.2013.796550

Rajendiran S, Coumar MV, Kundu S, Dotaniya AM, Rao AS (2012) Role of phytolith occluded carbon of crop plants for enhancing soil carbon sequestration in agro-ecosystems. Curr Sci 103 (8):911–920. https://www.jstor.org/stable/24088879

Rajendiran S, Coumar V, Dotaniya ML, Kumar A (2016) Carbon occlusion potential of rice phytoliths: implications for global carbon cycle and climate change mitigation. Appl Ecol Environ Res 14(2):265–281. https://doi.org/10.15666/aeer/1402_265281

Ramanan R, Kannan K, Sivanesan SD, Mudliar S, Kaur S, Tripathi AK, Chakrabarti T (2009) Bio-sequestration of carbon dioxide using carbonic anhydrase enzyme purified from *Citrobacter freundii*. World J Microbiol Biotechnol 25(6):981–987. https://doi.org/10.1007/s11274-009-9975-8

Rivadeneyra MA, Martín-Algarra A, Sánchez-Román M, Sánchez-Navas A, Martín-Ramos JD (2010) Amorphous ca-phosphate precursors for ca-carbonate biominerals mediated by *Chromohalobacter marismortui*. ISME J 4(7):922. https://doi.org/10.1038/ismej.2010.17

Roh Y, Zhang CL, Vali H, Lauf RJ, Zhou J, Phelps TJ (2003) Biogeochemical and environmental factors in Fe biomineralization: magnetite and siderite formation. Clays Clay Miner 51 (1):83–95. https://doi.org/10.1346/CCMN.2003.510110

Rubin E, De Coninck H (2005) IPCC special report on carbon dioxide capture and storage. TNO (2004): cost curves for CO_2 storage, Part 2. Cambridge University Press, Cambridge, p 14

Ruiz N, Ward D, Saltz D (2002) Calcium oxalate crystals in leaves of *Pancratium sickenbergeri*: constitutive or induced defence? Funct Ecol 16(1):99–105

Sadorsky P (2009) Renewable energy consumption, CO_2 emissions and oil prices in the G7 countries. Energy Econ 31(3):456–462. https://doi.org/10.1016/j.eneco.2008.12.010

Shaffer G (2010) Long-term effectiveness and consequences of carbon dioxide sequestration. Nat Geosci 3(7):464. https://doi.org/10.1038/ngeo896

Sharma A, Bhattacharya A (2010) Enhanced biomimetic sequestration of CO_2 into $CaCO_3$ using purified carbonic anhydrase from indigenous bacterial strains. J Mol Catal B: Enzym 67 (1–2):122–128. https://doi.org/10.1016/j.molcatb.2010.07.016

Smith KS, Ferry JG (2000) Prokaryotic carbonic anhydrases. FEMS Microbiol Rev 24(4):335–366. https://doi.org/10.1111/j.1574-6976.2000.tb00546.x

Snyder CS, Bruulsema TW, Jensen TL, Fixen PE (2009) Review of greenhouse gas emissions from crop production systems and fertilizer management effects. Agric Ecosyst Environ 133 (3–4):247–266. https://doi.org/10.1016/j.agee.2009.04.021

Tambutté S, Tambutté E, Zoccola D, Caminiti N, Lotto S, Moya A, Allemand D, Adkins J (2007) Characterization and role of carbonic anhydrase in the calcification process of the azooxanthellate coral *Tubastrea aurea*. Mar Biol 151(1):71–83. https://doi.org/10.1007/s00227-006-0452-8

Van Gemerden H (1993) Microbial mats: a joint venture. Mar Geol 113:3–25. https://doi.org/10.1016/0025-3227(93)90146-M

Verrecchia EP, Braissant O, Cailleau G (2006) The oxalate-carbonate pathway in soil carbon storage: the role of fungi and oxalotrophic bacteria. In: Gadd GM (ed) Fungi in biogeochemical cycles, vol 24. Cambridge University Press, Cambridge, p 289

Visscher PT, Prins RA, van Gemerden H (1992) Rates of sulfate reduction and thiosulfate consumption in a marine microbial mat. FEMS Microbiol Lett 86(4):283–293. https://doi.org/10.1111/j.1574-6968.1992.tb04820.x

Von Blottnitz H, Curran MA (2007) A review of assessments conducted on bio-ethanol as a transportation fuel from a net energy, greenhouse gas, and environmental life cycle perspective. J Clean Prod 15(7):607–619. https://doi.org/10.1016/j.jclepro.2006.03.002

Wigand M, Kaszuba JP, Carey JW, Hollis WK (2009) Geochemical effects of CO_2 sequestration on fractured wellbore cement at the cement/caprock interface. Chem Geol 265(1–2):122–133. https://doi.org/10.1016/j.chemgeo.2009.04.008

Yadav R, Labhsetwar N, Kotwal S, Rayalu S (2011) Single enzyme nanoparticle for biomimetic CO_2 sequestration. J Nanopart Res 13(1):263–271. https://doi.org/10.1007/s11051-010-0026-z

Yoshida N, Higashimura E, Saeki Y (2010) Catalytic biomineralization of fluorescent calcite by the thermophilic bacterium *Geobacillus thermoglucosidasius*. Appl Environ Microbiol 76(21):7322–7327. https://doi.org/10.1128/AEM.01767-10

Chapter 6
A Review of Coupled Geo-Chemo-Mechanical Impacts of CO_2-Shale Interaction on Enhanced Shale Gas Recovery

Danqing Liu, Sen Yang, Yilian Li, and Ramesh Agarwal

Abstract Shale is an important geological media for carbon capture utilization and storage. On one hand it can be regarded as impermeable caprock to prevent CO_2 migration from reservoir, and on the other hand it can be also treated as natural gas and CO_2 storage reservoir. CO_2-shale reactions within caprock can interfere the rock integrity so as to bring threatens to the long-term carbon storage safety and stability, however this interaction can also improve rock conductivity to enhance shale gas recovery for the organic-rich shale. This article presents a review of the current state of knowledge regarding CO_2 and shale interactions and their potential impacts on shale properties and groundwater quality in the context of CO_2 enhanced shale gas recovery. The characterization of gas shale and CO_2, which is critical to the occurring of different interactions between CO_2 and shale, is firstly summarized. The major interaction mechanisms between CO_2 and shale including CO_2-shale-water geochemical reactions, CO_2 adsorption induced clay swelling and organic matter extraction with supercritical CO_2 and their impact on rock porosity, permeability, mechanical properties, gas adsorption capacity and groundwater quality are surveyed. Finally, the open questions in the field are emphasized and new research needs are highlighted.

Keywords CO_2 enhanced shale gas recovery · CO_2-shale interaction · Porosity and permeability · Mechanical properties · Adsorption capacity · Groundwater quality

---

D. Liu
State Key Laboratory of Biogeology and Environmental Geology, China University of Geosciences, Wuhan, China

School of Environmental Studies, China University of Geosciences, Wuhan, China

S. Yang · Y. Li
School of Environmental Studies, China University of Geosciences, Wuhan, China

R. Agarwal (✉)
Washington University in St. Louis, St. Louis, MO, USA
e-mail: rka@wustl.edu

Inamuddin et al. (eds.), *Sustainable Agriculture Reviews 37*, Sustainable Agriculture Reviews 37, https://doi.org/10.1007/978-3-030-29298-0_6

6.1 Introduction

Carbon capture utilization and storage, as a critical tool in addressing the global climate change and energy shortage crisis, can not only significantly reduce emissions from power generation but also provide deep emission reduction strategy for key industrial processes including steel, cement and manufacturing (International Energy Agency). Shale is a kind of low-permeability, low porosity rock with great potential containing high abundance of oil/gas reserves in reducing sedimentary environment. It is widely distributed all around the world and is an important geological media for carbon capture utilization and storage.

Shale can be regarded as caprock for long-term CO_2 storage due to its merits of low hydraulic conductivity and preventing CO_2 leakage from reservoir. However, the organic-rich shale can be also treated as CO_2 storage reservoir mainly benefits from its high adsorption affinity to CO_2 over CH_4 (Fig. 6.1). In the meantime of CO_2 storage in shale reservoir by adsorption, CH_4 is replaced and desorbed from the shale matrix and migrates towards to the production well, and the shale gas recovery ratio can be simultaneously enhanced. Furthermore, the supercritical CO_2 or liquid CO_2 can be used as fracking fluid. Take advantage of its low viscosity, high diffusivity

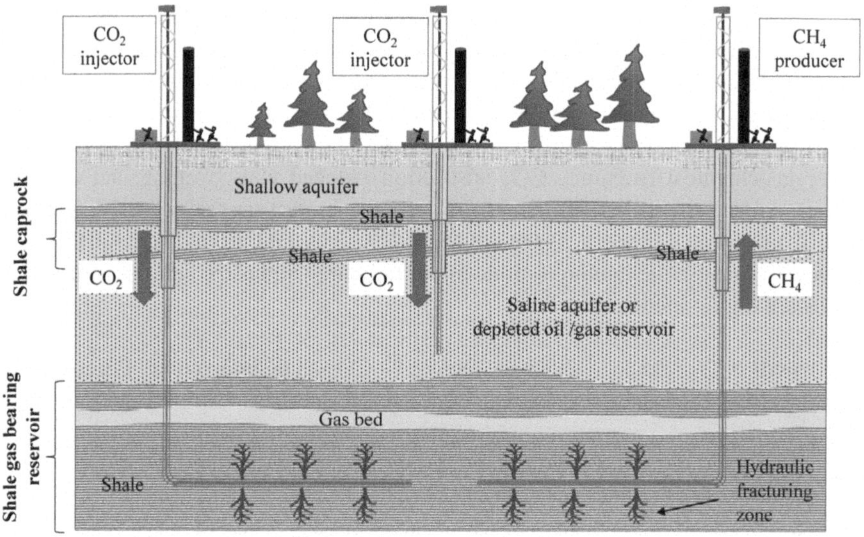

Fig. 6.1 The utilization of shale in the context of CO_2 capture, storage and utilization. Shale, as low permeable rock, is conventionally regarded as the caprock for CO_2 storage in deep saline aquifer or depleted oil/gas reservoir. In shale gas bearing reservoir, CO_2 can be injected to stimulate the reservoir as well as to store for permanent

and anhydrous property, supercritical CO_2/liquid CO_2 fracturing is even more superior to hydraulic fracturing in shale reservoir stimulation. Owing to the win-win advantages of reducing CO_2 emission through permanent storage as well as enhancing shale gas recovery, CO_2 enhanced shale gas recovery technology attracts more and more attentions in recent years.

It is well known that the interaction between CO_2 and caprock can interfere the caprocks integrity and bring threatens to the long-term carbon storage security and stability. Reactions among CO_2, brine, and caprocks and associated impacts on flow paths and hydraulic rock properties have been researched extensively. For example, Skukla et al. (2010) comprehensively reviewed the impact of CO_2 injection on caprock integrity mainly took into consideration of geophysical aspects. The stability response of the caprock during and after CO_2 injection were discussed, while the impact of pre-existing fractures and re-opened cracks was also systematically analyzed. Griffith et al. (2011) presented another comprehensive review on caprocks across the U.S. considered for pilot and large-scale demonstration project of CO_2 capture and storage, they focused mainly on structural, petro-physical, and chemical characteristics. Liu et al. (2012) also conducted a short review on the geochemical aspects of the injected CO_2 on caprock integrity.

The efficiency of the fracking process is dependent on the retention of permeability generated through fracturing. Reaction of the host shale rock with the fracking fluids or brine can also cause the formation of new phases (such as swelling clays and precipitated minerals). These new phases can block porosity and lead to sudden reductions in production rates. The shale gas enhancing efficiency with CO_2 injection is mainly controlled by the rock preferential adsorption capacity to CO_2 over CH_4. The CO_2 interaction with rock can impact the organic geochemical properties of shale so as to interfere its adsorption capacity to different gases. To sum up, CO_2-shale interaction is also critical for CO_2 enhanced shale gas recovery technology. However, the coupled geochemical-mechanical-adsorption properties change responding to CO_2 injection in the organic-rich gas shale has received less attention.

This paper review the current state of knowledge regarding CO_2 and shale interactions and their potential impacts on shale properties and groundwater quality in the context of CO_2 enhanced shale gas recovery. Firstly, we summarize the characterization of gas shale and CO_2, which is critical to the occurring of different interactions between CO_2 and shale. Then the major three mechanisms including CO_2-shale-water geochemical reactions, CO_2 adsorption induced clay expansion and organic matter extracting with supercritical CO_2 were discussed in detail. Afterwards, the impact of the aforementioned effects on shale properties such as porosity, permeability, mechanical properties and gas adsorption capacity was surveyed. Finally, we also highlight the present research progress of the effect of CO_2-shale interaction on groundwater quality.

6.2 Properties of Shale and CO_2

6.2.1 Shale

Shale is a typical sedimentary rock composed mainly of organic matter and inorganic matters. Among which, the organic matter indicates kerogen and asphaltene which is also regarded as the major adsorbing media for the adsorption gas in the shale reservoir (Zhao et al. 2017; Xiong et al. 2017). As for the inorganic matter, clay mineral, quartz, calcite and feldspar all play important roles (Liang et al. 2014; Yang et al. 2015). The total organic carbon content is one of the critical factor controlling the gas adsorption ability of shale. In the meantime, the mineral composition, kerogen type of shale also significantly related to the gas containing potential of shale (Ross and Bustin 2009).

Extremely low permeability and porosity is another prominent characteristic of gas shale, which decides the shale gas reservoir to be the kind of continuous natural gas reservoir combines generation, storage and cover. The particle size of shale reservoir is quite small and it is mainly made up of silt with size <5 μm and mud or sand with a size between 5 μm and 60 μm (Tan et al. 2014; Mokhtari and Tutuncu 2015; He et al. 2017). It is reported that the average porosity of shale is less than 10%. The average porosity of shale in the USA is between 2.5% and 6.6%, while the average porosity of marine shale from the Southern China is less than 4% (Tan et al. 2014; Chalmers et al. 2012). The permeability of shale is also quite low which varies within the range of nano-Darcy and two orders of magnitude lower than the tight sandstone reservoir. It is investigated that the permeability of marine shale from the Southern China is less than 20.2 nano-Darcy (Tan et al. 2014).

The pore structure of shale is the deciding factor of shale gas adsorption properties (Clarkson et al. 2013; Tian et al. 2015; Jiang et al. 2016a, b; Liang et al. 2017; Song et al. 2018). Based on the size of pore diameter, the pore in shale can be divided into three types according to the International Union of Pure and Applied Chemistry (IUPAC): 1. Micropore, diameter < 2 nm; 2. Mesopore, 2 < diameter < 50 nm; 3. Macropore, diameter > 50 nm (Li et al. 2015). The pore in shale can be also classified as mineral pore, organic pore and fracture-typed pore according to its morphology and location (as shown in Fig. 6.2). It is reported that the presence of a large number of organic nanoscale micro-pores is the main controlling factors of shale gas accumulation.

6.2.2 CO_2/Supercritical CO_2

As a colorless and odorless gas, the carbon dioxide (CO_2) is ubiquitous in nature, which content is about 0.03–0.04% in the atmosphere. However, the content of CO_2

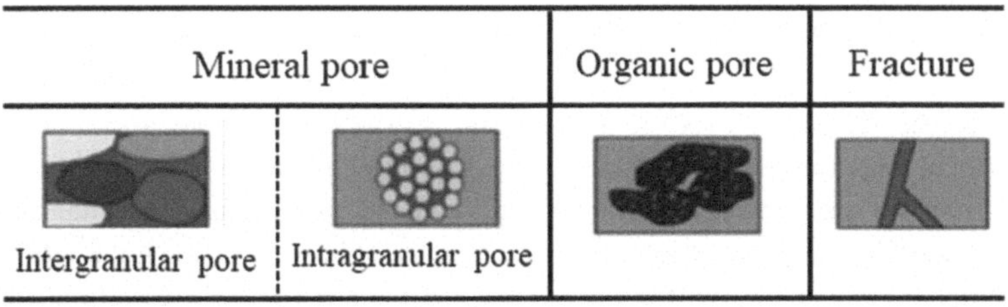

Fig. 6.2 Pore-type of shale. There are mainly three types of pores in shale, which including the mineral pore (can be divided as intergranular pore and intragranular pore), organic pore and fracute-type pore

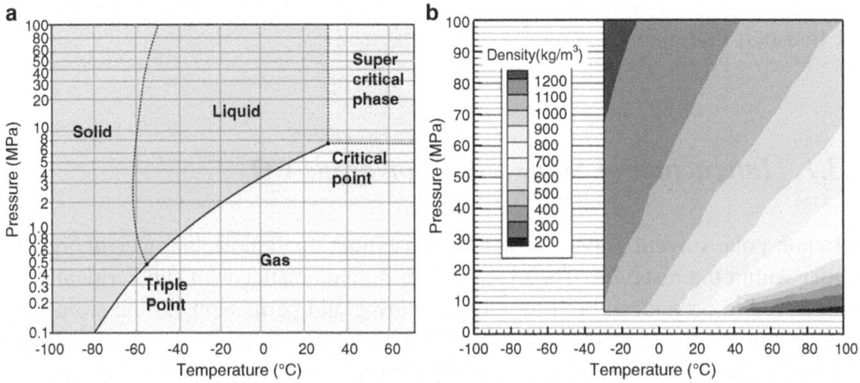

Fig. 6.3 (**a**) CO_2 phase diagram. The horizontal axis is temperature (°C) and y axis is pressure (Mpa), the critical point of CO_2 transfer from liquid or gaseous phase to supercritical phase is at the temperature of 31.1 °C, pressure of 7.38 MPa. (**b**) CO_2 density evolution with temperature and pressure. The x axis represents temperature (°C) while the y axis is pressure (MPa). CO_2 density is large at low temperature and high pressure but low at high temperature and low pressure

in atmosphere increases gradually with the development of industry, resulting in the global greenhouse effect. The nontoxicity, noninflammability and gaseous state of CO_2 makes it widely used in many industrial fields. With the variation of temperature and pressure, the phase state of CO_2 is quite different (as shown in Fig. 6.3).

CO_2 exists as supercritical phase when the ambient temperature and pressure exceeds its critical value of 31.1 °C and 7.38 MPa. Different from either liquid CO_2 or gaseous CO_2, the supercritical CO_2 has low viscosity and high diffusivity like gas, as well as similar density to liquid CO_2 and has ability to dissolve certain substances (Zhou et al. 2003). The surface tension of supercritical CO_2 is too weak to close to zero, which enables supercritical CO_2 to penetrate easily into the microstructure in shale. What's more, supercritical CO_2 is also a non-polar solvent. It can extract lipids, volatile mater, and some smaller molecular weight substance within the reservoir.

6.3 Interaction of CO_2 and Shale

Owing to the reactivity of CO_2 and special properties of shale, the injection of supercritical CO_2 into the gas-bearing shale can induce extensive reactions not only on minerals but also on organic matters and undoubtedly impacts the shale properties. Injected CO_2 is present as supercritical or gas phase in shale for fracturing or long-term storage. Then it will dissolve into pore water and creating carbonic acid to interfere the mineral reaction equilibrium. The supercritical CO_2 can extract organic matters to reshape the organic microfracture of shale and affects its adsorption capacity. Meanwhile, molecular CO_2 can adsorb from the supercritical or water phase onto clay surfaces within the interlayer spaces and can displace water. This process can result in shrinkage, altering the inter-granular pore space and possibly the transport flow paths and mechanical properties.

6.3.1 Interaction of Shale with Anhydrous CO_2

As a non-polar solvent, supercritical CO_2 can extract small molecular weight organic matter within the reservoir. The extraction of organic matter with supercritical CO_2 is deserved to be noticed both for CO_2 fracturing and permanent carbon storage in gas shale reservoir. Supercritical CO_2 extracting of organic matter has been systematically studied in the field of chemical engineering (Naik et al. 1989), its application into geological principle started from the extracting of organic matter from oil shale (Bondar and Koel 1998; Akinlua et al. 2008; Allawzi et al. 2011). Bondar and Koel (1998) used the supercritical CO_2 to effectively extract hydrocarbon from oil shale, and found the yield depended on the inorganic matrix. In order to provide a more accurate indication of hydrocarbon yield from supercritical CO_2 extraction, Jarboe et al. (2015) documented the extractable organic matter to total organic carbon content of shale after supercritical CO_2 treatment. Their results showed the yield was as a function of sample matrix size, and the supercritical CO_2 extracted diesel-range n-aliphatic hydrocarbons from high-maturity shale at reservoir conditions. Jin et al. (2017) carried out the hydrocarbons extraction from Bakken shale at reservoir conditions, and found 15–65% hydrocarbon was extracted by supercritical CO_2. In addition, Wu et al. (2015) carried out molecular dynamics simulations to focus on the effect of supercritical CO_2 on the dissolution of kerogen from oil shale, and observed that the supercritical CO_2 dissolving capacity increasing with pressure as well as temperature at high pressure.

Recently, some researchers also studied the supercritical CO_2 extracting effect on organic matter during CO_2 geological storage. Scherf et al. (2011) investigated the type and content of organic matter extracted by supercritical CO_2 in the process of carbon storage in deep saline aquifer and results showed that the polar lipid fatty acids and low molecular weight organic acids were extracted. Besides, rock porosity, permeability and mineral composition might impact the organic extracting efficiency. Kolak and Burruss (2006) explored the supercritical CO_2 extracting effect in different rank coals and mentioned that supercritical CO_2 has a high extraction capacity for alkanes and polyaromatics, and the extraction amount is related to coal rank. Okamoto et al. (2005) also reveal that the CO_2 extraction of organic matter can reshape the pore structure of caprock under supercritical condition, which may bring threatens to CO_2 storage security and stability.

However, the extraction of organic matter by supercritical CO_2 plays positive role in the process of CO_2 fracturing. Jiang et al. (2016a, b) ascribed the shale gas seepage channels widened during CO_2 injection to the extraction of organic matter by supercritical CO_2. Zhang et al. (2017a) also observed the promotion of seepage-flow pores in shale, the significantly decreasing of compounds with weakly polar functional groups and seldom changing of the strong polar functional groups of shale after treatment of supercritical CO_2. Pan et al. (2018a, b) compares the interactions between supercritical CO_2 and two shale samples from the marine Longmaxi formation and terrestrial Yanchang formation, their results indicated drastic variation of shale nano-pore structure to supercritical CO_2 treatment for both of the two shale samples. They claimed that chemical reactions in minerals, the swelling in clay and supercritical CO_2-induced extraction of hydrocarbons were the major cause for this phenomenon. Organic pores are reported to be the major conduits for oil and gas within shale (Thomas and Clouse 1990). Consequently, the extraction of organic matter with supercritical CO_2 can be helpful for the generation of organic pores in shale and promote the flowing of oil/gas towards production well.

6.3.2 CO_2-Water-Rock Geochemical Reactions in Shale

As one of the water-soluble gas, the dissolution of CO_2 into formation water can result in its acidification, and accelerates the geochemical reaction during geothermal exploitation as well as CO_2 sequestration (Liu et al. 2016; Cui et al. 2018). With the development of CO_2 enhanced shale gas recovery technology, much more studies have been carried out to focus on the CO_2-water-rock interaction in shale. Alemu et al. (2011) carried out experiment to study the interaction between CO_2, brine and shale including carbonate-rich and clay-rich shale. Their results showed that carbonate-rich shale was more reactive than clay-rich shale, and the temperature had no significant effect on mineralogical changes of clay-rich shale. The high reactivity of carbonate was consistent with previous study about supercritical CO_2-sandstone interaction, ascribing to the fast dissolution kinetics of calcite and dolomite. Study of carbonate-rich shale with CO_2-charged brine by Ilgen et al. (2018) provided a powerful evidence for carbonate dissolution.

Despite of carbonate, the clay mineral such as kaolinite, illite, chlorite and quartz as well as feldspar can also dissolve at weak acidic environment. For example, Liu et al. (2012) found the dissolution of feldspar and anhydrite as well as the precipitation of siderite, illite and smectite during shale-brine-CO_2 interaction at 200 °C and 300 bars. Similarly, Rezaee et al. (2017) documented the dissolution of potassium feldspar and reduction of illite-smectite mixed layer during shale-brine-CO_2 at 66 °C and 151.7 bar. Luo et al. (2019) also found the contents of clay minerals and carbonate minerals decrease when reacted with CO_2, whereas the contents of siliceous minerals increase in the system.

6.3.3 CO_2 Adsorption Induced Swelling in Shale

The clay minerals such as smectite, kaolinite, chlorite and illite that made up of shale are reported to have high capacity of sorption due to their wide internal surface area. The naturally occurring clay minerals consist predominantly of stacks of two-dimensional aluminosilicate sheets, which carry a negative charge (Chen et al. 2008; Anderson, 2010), and a interlayer space with widths in a nanometer scales contains positively charged cations (Fig. 6.4). The cations in the interlayer determine the likelihood of intercalation or exfoliations of molecules into or out of the clay's

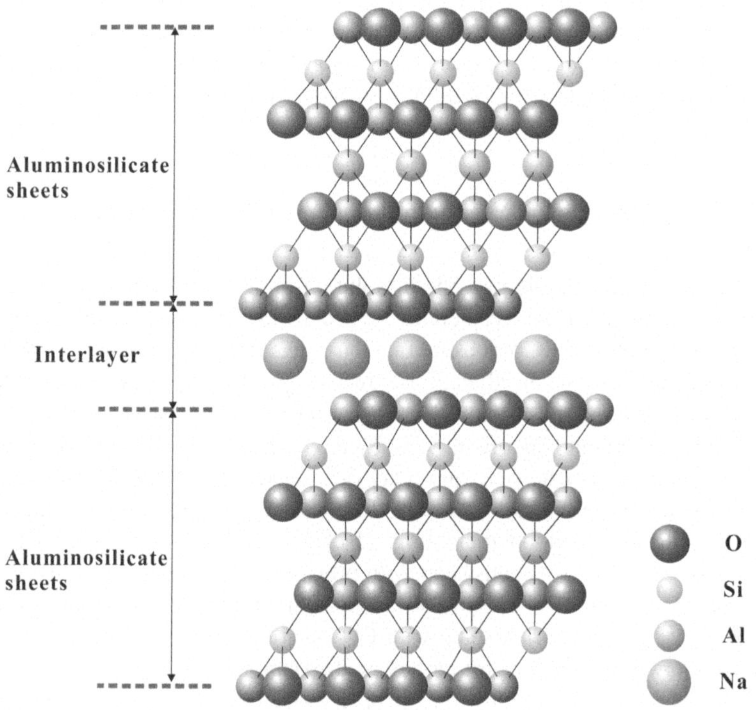

Fig. 6.4 Layered structure and basal unit cell dimension of smectite clay minerals. The aluminosilicate sheets include tetrahedral and octahedral structure and are mainly composed of Al^{3+}, Mg^{2+}, Fe^{3+} and Si^{4+} oxides. Between two aluminosilicate layers, there is an interlayer contained exchangeable species like Na^+, K^+, Ca^{2+}

lattice and also cause clays to swell (hydration) or to collapse (dehydration) (Chen et al. 2008).

Several studies to date have reported measurements of swelling in clay minerals induced by CO_2 adsorption. The diffusive transport and gas sorption experiments of shale and clay minerals showed that the high CO_2 adsorption capacity of shale was attributed to the clay minerals (Busch et al. 2008). Based on shale from New Albany, Lahann et al. (2013) discovered the phenomenon of swelling in CO_2 adsorption experiments, but the expansion process has not been qualitatively described. Heller and Zoback (2014) measured the adsorption swelling of carbon, kaolinite and illite, and studied the relationship between gas adsorption capacity and the swelling strain of minerals. Schaef et al. (2015) for example, also investigated the interaction of variably hydrated supercritical CO_2 and Na-, Ca-, and Mg-exchanged MMT. Structural volumetric changes were observed, and results indicated that intercalation of CO_2 was inhibited without the participating of water. The introducing of a little water and the partial expansion of the interlayer region can significantly increases the intercalation of CO_2, however it will then gradually decrease with further hydration of clay. Similar observations were also made by Giesting et al. (2012) and Busch et al. (2008).

Quite many factors impact the swelling effect induced by CO_2 adsorption. Lu et al. (2016) found that the maximum volumetric swelling of shale decreases with temperature, and CO_2-induced adsorption is dominant under low pressure, while the volumetric swelling under high pressure is mainly affected by gas pressure. Chen et al. (2018) discussed the relationship of adsorption-induced swelling strain and gas pressure by conducting the shale deformation experiment with helium and methane under constant confining pressure and different gas pressure. They also measured the volumetric strain process of organic-rich shale within various gas adsorption, and found that the volumetric strain first rose rapidly and then stabilized, which was related to the gas occupying cracks and macropore in the initial stage and filling micropore in the later stage. Bakhshian and Hosseini (2019) built a comprehensive model to study the effect of temperature and pores geometry on CO_2 adsorption of shale. They found that the faster and higher amount of CO_2 adsorption on duct pore than slit pore was ascribed to the large attractive surface field. Besides, the pore shape as well as temperature had more significant effect on the volumetric strain of mesopores than micropore.

6.4 Effect of CO_2-Shale Interaction on Rock Properties

6.4.1 Porosity and Permeability

Micro-structure of shale can be influenced by CO_2-shale interactions through mineral dissolution and precipitation, adsorption-induced expansion effects as well as organic matter extraction with supercritical CO_2 (Fig. 6.5). Jiang et al. (2016a, b) indicated that the extraction of organic matte by supercritical CO_2 was responsible

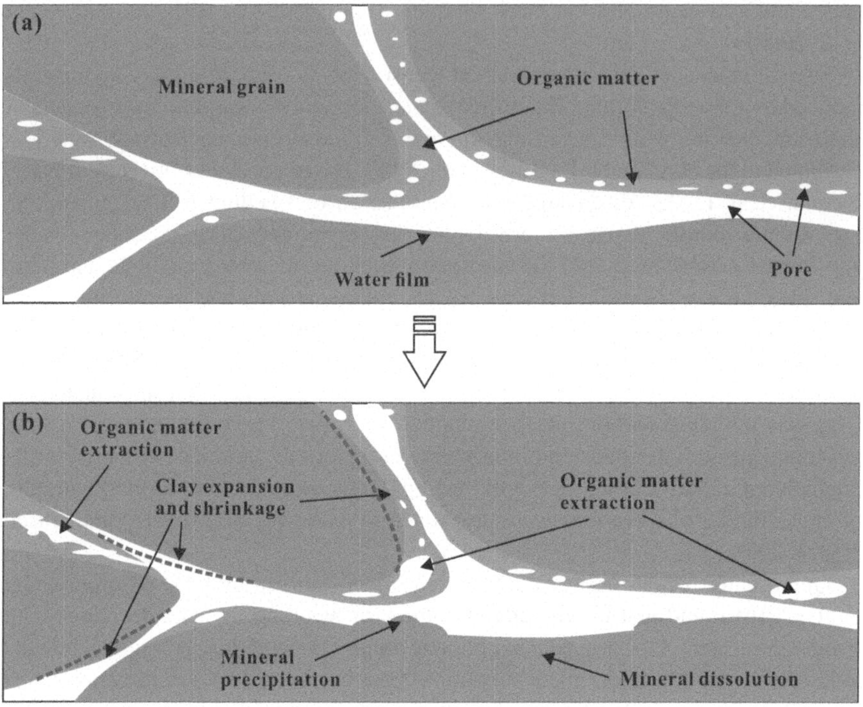

Fig. 6.5 Schematic of the three possible mechanisms responsible for shale micro-structure variation (revised from Pan et al. 2018b). (**a**) stands for Raw sample and (**b**) represents supercritical CO_2-treated sample. The mechanism of organic matter extraction with supercritical CO_2, clay expansion and shrinkage induced by CO_2 adsorption and also the mineral dissolution and precipitation cause the physical-chemo-mechanical property variation of shale during CO_2 injection

for shale pore volume and specific surface area increasing, and the average porosity enhanced with the increasing of temperature and pressure were 125.93% and 81.48%, respectively. Similarly, Pan et al. (2018b) mentioned that the dissolving extraction of organic matter from Longmaxi shale could lead to the reduction of micropores and fine mesopores. When the ambient temperature and pressure is below the critical point of CO_2, the supercritical CO_2 can transform into subcritical CO_2, resulting in the disparate variation of shale pore structures (Span and Wagner 1996). Pan et al. (2018a) compared the shale pore structures evolution after subcritical and supercritical CO_2 treatment, and concluded that different states of CO_2 induce different variation tendency of shale pore structure. When compared with interaction of shale and subcritical CO_2, interaction of shale with supercritical CO_2 can creates more drastic effect on pore structure. Pan et al. attribute this phenomenon to the greater dissolution, expansion effect and the extraction mechanism associated with supercritical CO_2.

In addition to organic matter extraction with dry supercritical CO_2, mineral dissolution and precipitation can also happen when shale contacts with anhydrous

CO_2. Microspore reduction after dry supercritical CO_2 treatment was found by Yin et al. (2016) and Sanguinito et al. (2018). However, in real environment, the injected CO_2 cannot fully displace fluids in shale reservoir and water always exists in the pores. Researches indicate that the shale-CO_2 interaction can be more intensive as water involved in the CO_2-shale system. For example, Goodman et al. (2019) found that the microspores can be completely dissolved during shale-water-CO_2 interaction, and the larger generated pore volume after reaction compared with anhydrous CO_2 system was attributed mainly to the rapid and drastic carbonate minerals dissolution in the participate of water. Furthermore, different mineral composition of shale will result in various shale-CO_2 interaction type, leading to the discrepant pore structures evolution of shale. Zou et al. (2018) found etched pores and cavities at the surface of shale after shale-water-CO_2 interaction, and the mineral composition difference contributed to the various porosity changes. However, larger dissolved pore of carbonate-poor shale than that of carbonate-rich shale was found after the 168 h treatment, which was related to the high initial porosity and permeability of shale resulting in the easy contact with shale (Yu et al. 2012).

As a key parameter to evaluate the efficiency of gas flow ability, the permeability is related to change of pore structures after CO_2-shale interaction. Studies have shown that the dissolution of minerals can enlarge the pore volume and increase rock permeability (Deng et al. 2015; Soong et al. 2017; Zou et al. 2018). However, it has been found that permeability decline was also ascribed to the secondary mineral precipitation as well as pore roar blockage by clay minerals (Yu et al. 2012; Jones and Detwiler 2016). Besides, the asphaltene precipitating and deposition in shale severely decreased its permeability during CO_2 huff and puff injection (Shen and Sheng 2017). The complex composition and pressure change as well as the injection CO_2 concentration all influences the asphaltenes stability. According to the experiment of Shen and Sheng (2018), 48.5% permeability reduction occurred after 6 cycles of CO_2 huff-puff injection within 26.8% permeability reduction at the first cycle.

6.4.2 *Mechanical Properties*

The mechanical properties of shale after exposure with CO_2 is essential for shale gas production and long-term CO_2 sequestration. As mentioned previously, the injected CO_2 can dissolve into water to from carbonic acid, which has the ability to dissolve mineral of shale, change shale structure and then affect its mechanical properties. For example, Lyu et al. (2016) measured the uniaxial compressive strength and Young's modulus variation of shale exposure with gaseous and supercritical CO_2, and found the reduction of 56.43% and 66.05% in uniaxial compressive strength, as well as 54.21% and 56.32% in Young's modulus, respectively, they claims the reason to the mineral dissolution and carbonate precipitation. Zhang et al. (2017b) also found that the concentration reduction of Fe, K, Mg, Na, Al, and especially for Ca in CO_2 saturated and CO_2-brine saturated shale, accompanying with a large decrease of strength and elastic modulus, which indicated the relationship between minerals variation and mechanical behaviour of shale. In addition, the crystal water in shale can combine with CO_2 to create acid environment, facilitating the shale-CO_2 interaction and influencing the mechanical properties of shale (Liu et al. 2017).

Apart from mineral dissolution, the CO_2 adsorption can decrease the surface potential energy of shale and resulting its swelling. The supercritical CO_2 was more remarkable than subcritical CO_2 for strength and elastic modulus reduction of shale, ascribing to the pressure difference for CO_2 permeation and CO_2 adsorption as well as lower viscosity and no capillary force of supercritical CO_2 (Lyu et al. 2018). To distinguish the adsorption and gas pressure effects on CO_2-induced swelling of shale, Ao et al. (2017) carried out the strain tests of shale with supercritical CO_2 and no-adsorption gas (He), and indicated that CO_2 adsorption was responsible for shale deformation at low pressure, while gas pressure resulted in shale deformation at high pressure. Feng et al. (2019) discussed the relationship between failure behavior and bedding angle of shale after supercritical CO_2 adsorption and indicated that the layer direction resulted in the tensile failure as well as shear failure of shale.

Since the organic matter in shale can be extracted by supercritical CO_2, the pores dissolution after extraction results in variation of shale mechanical properties. When interacted with organic rich shale, Yin et al. (2017) noticed that the uniaxial compressive strength and elastic modulus decreased more significantly after supercritical CO_2 exposure than subcritical CO_2 exposure. Similar results were also found by Lu et al. (2019) when exposed the Longmaxi shale with gaseous and supercritical CO_2. This difference demonstrated that pore structures variation of shale was not only affected by minerals dissolution and precipitation as well as adsorption-induced swelling of CO_2, but also ascribed to the organic matter extraction by supercritical CO_2. The combination of the above mentioned three mechanism resulted in the mechanical properties changes of shale after shale-CO_2 interaction.

6.4.3 Adsorption Properties

Abundant researches indicated the CO_2 adsorption capacity of shale was stronger than CH_4, which was the basis for the development of CO_2 enhanced shale gas recovery technology (Nuttall et al. 2005; Chareonsuppanimit et al. 2012; Duan et al. 2016; Huo et al. 2017; Yang et al. 2018). There is no doubt that the pore structures as well as the minerals composition will change after CO_2 injection, which affecting the gas adsorption capacity of shale. However, limited research has been carried out to focus on the adsorption capacity variation. Zhou et al. (2018) documented that the adsorption capacity of shale was mainly controlled by the micro- and mesopores, and the organic matter extraction as well as adsorption-induced expansion resulted in the decrease of shale adsorption capacity of CO_2 and CH_4 after supercritical CO_2 treatment. Similar results obtained by Hui et al. (2019), and they also ascribed the specific surface area variation to extraction and adsorption-induced expansion rather than mineral dissolution during shale- supercritical CO_2 interaction. However, the distinct phenomenon that terrestrial Yanchang shale showed a decrease of adsorption capacity with specific surface area increase after supercritical CO_2 exposure indicating that the reduction of oxygen-containing species was responsible for the decrease of CO_2 maximum adsorption in shale. The discrepant adsorption behavior of shale after CO_2 exposure manifest that more studies are needed to clarify this problems. Furthermore, the effect of mineral dissolution and precipitation on shale adsorption behavior after CO_2 exposure is neglected by current research due to the limited interaction period. However, rapid and significant geochemical reactions exist within the timescale of shale gas recovery, and the geochemical effect on long-term CO_2 storage is proved to be potential. As a result, this problem is also an important issue needed to be resolved.

6.5 Effect of CO_2-Shale Interaction on Groundwater Quality

The effect of CO_2-water-rock interaction on element migration of shallow aquifers and saline formation have been investigated widely through field monitoring, laboratory experiment as well as geochemical modeling (Viswanathan et al. 2012; Romanak et al. 2012; Rillard et al. 2014; Rathnaweera et al. 2016). Researches demonstrated that the injection of CO_2 can dissolve specific minerals such as calcite and dolomite, resulting in the groundwater contamination.

Previous studies showed that trace elements had various affinities for different minerals and organic matter in shale (Glikson, et al. 1985; Ripley et al. 1990; Phan et al. 2015). When interacted with CO_2 and brine, the various minerals dissolution results in the trace elements release at different extent. For example, Jean et al. (2015) carried out the shale-CO_2-water interaction at 25 MPa and 90 °Cfor 7 days, and found large amount of V, Cr, Co, Cu, and Rb dissolved from shale, while the dissolution of Zn, Se, Mo, and Cd was minimal. Marcon and Kaszuba (2015) indicated that the release content of Cd, Co, Cu, Cr, Fe, Ni, Pb, V, and Zn in the mixture of limestone and shale were higher than that of limestone only. Similarly, Luo et al. (2019) measured the element mobilization of shale before and after shale-CO_2-water interaction, and found the mobility of the trace elements was relatively low when compared with major elements. In addition, the study also indicated that the elements release was not only ascribed to the mineral dissolution, but also related to the cation exchange.

Many organic compounds such as benzene, toluene and naphthalene had been found in produced water or flowback fluids (Strong et al. 2013; Ziemkiewicz, 2013; Cluff et al. 2014; Akob et al. 2015), demonstrating the organic matter in shale becoming another potential source of groundwater contamination. To clarify the possible contamination of organic matter in shale, Dustin et al. (2018) isolated kerogen from both Green River and Marcellus shales, and conducted the experiment of kerogen-fracturing fluids interaction. Their study showed that organic matter dissolved in both fracturing fluid and hydrochloric acid solution (pH = 2), and dissolution of kerogen contributed to the release of heavy metals. Although the functional group alteration of kerogen had been noticed in this study, analysis of organic components in solution was not involved. Moreover, the supercritical CO_2 can effectively extract organic matter in shale, which also means its potential release during shale-water-CO_2 interaction. Thus, the potential organic contamination during CO_2-water-rock geochemical reactions in shale needs further in-depth study.

6.6 Conclusions

Gas shale is mainly composed of clay minerals, quartz, feldspar and plenty of organic matter. The clay mineral is deemed with sorption ability and can expand or contract when contacting with CO_2 through adsorption; the organic matter can also be extracted by supercritical CO_2; while the geochemical reactions can also occur between shale, water and CO_2. Thus, it can be seen that the characteristics of CO_2 and shale determines that a series of interaction will inevitably occur once CO_2 contacting with shale. In addition, as a low-porosity and low-permeability reservoir, tiny influence on the conductivity of shale may exert huge impact on gas recovery and CO_2 storage efficiency. As a result, the interaction between CO_2 and shale is critical for the context of CO_2 enhanced shale gas recovery.

CO_2-water-shale geochemical reactions can cause minerals dissolution and expansion, CO_2 adsorption onto clay will induce the swelling of minerals, and the extracting of organic matter with supercritical CO_2 may lead to the organic dissolution. The aforementioned mechanisms all have potential to interfere the porosity, permeability, mechanical properties and even the adsorption capacity of shale. However, it is difficult to distinguish these three mechanisms in the current study, and it is impossible to quantify the influence weight of different mechanisms on shale property variation. Advanced interdisciplinary methods are necessary to integrate the physical, mechanical and geochemical changes in order to predict the response at the pore, core, and reservoir scale.

The impact of CO_2-shale interaction on gas adsorption properties of shale is needed to be clarified. The effect of mineral dissolution and precipitation on shale adsorption behavior after CO_2 exposure is always neglected by current research due to the limited interaction period. However, rapid and significant geochemical reactions exist within the timescale of shale gas recovery, and the geochemical effect on long-term CO_2 storage is proved to be potential. As a result, this problem is also an important issue needed to be resolved. What's more, the CO_2-interaction induced organic contamination in the groundwater is also a critical problem waited to be explored.

The ultimate goal of studying the influence of CO_2-shale interaction on rock properties and groundwater quality is to assess the feasibility and potential of CO_2 enhanced shale gas recovery by considering the shale gas recovery efficiency, CO_2 storage capacity, stability and security as well as its environmental implication.

However, the recent researches mainly focused on lab-scale experiments and explored the micro-structure evolution of shale treated with CO_2. The relation between this variation and shale gas recovery and CO_2 storage was not built and it is difficult to provide straightforward conclusions for policy makers and stack holders. In the future, the coupled organic-geo-chemo-mechanical model should be established and field-scale experiments should be conducted to comprehensively explore the effect of CO_2-shale interaction on enhanced shale gas recovery.

Acknowledgments This work has been supported by the National Natural Science Foundation of China (NSFC, No. 41572233, No. 41902253), the China Postdoctoral Science Foundation funded project (No. 2018 M632943) has also provided partial support for this study.

References

Akinlua A, Torto N, Ajayi TR et al (2008) Supercritical fluid extraction of aliphatic hydrocarbons from Niger Delta sedimentary rock[J]. J Supercrit Fluids 45:57–63. https://doi.org/10.1016/j.supflu.2007.11.016

Akob D, Cozzarelli I, Dunlap D et al (2015) Organic and inorganic composition and microbiology of produced waters from Pennsylvania shale gas wells[J]. Appl Geochem 60:116–125. https://doi.org/10.1016/j.apgeochem.2015.04.011

Alemu BL, Aagaard P, Munz IA et al (2011) Caprock interaction with CO_2: a laboratory study of reactivity of shale with supercritical CO_2 and brine[J]. Appl Geochem 26:1975–1989. https://doi.org/10.1016/j.apgeochem.2011.06.028

Allawzi M, Al-Otoom A, Allaboun H et al (2011) CO_2 supercritical fluid extraction of Jordanian oil shale utilizing different co-solvents[J]. Fuel Process Technol 92:2016–2023. https://doi.org/10.1016/j.fuproc.2011.06.001

Anderson R, Ratcliffe I, Greenwell H et al (2010) Clay swelling—A challenge in the oilfield [J]. Earth Sci Rev 98:201–216. https://doi.org/10.1016/j.earscirev.2009.11.003

Ao X, Lu Y, Tang J et al (2017) Investigation on the physics structure and chemical properties of the shale treated by supercritical CO_2[J]. J CO2 Util 20:274–281. https://doi.org/10.1016/j.jcou.2017.05.028

Bakhshian S, Hosseini SA (2019) Prediction of CO_2 adsorption-induced deformation in shale nanopores[J]. Fuel 241:767–776. https://doi.org/10.1016/j.fuel.2018.12.095

Bondar E, Koel M (1998) Application of supercritical fluid extraction to organic geochemical studies of oil shales[J]. Fuel 77:211–213. https://doi.org/10.1016/S0016-2361(97)00188-9

Busch A, Alles S, Gensterblum Y et al (2008) Carbon dioxide storage potential of shales[J]. Int J Greenh Gas Control 2:297–308. https://doi.org/10.1016/j.ijggc.2008.03.003

Chalmers GR, Bustin RM, Power IM (2012) Characterization of gas shale pore systems by porosimetry, pycnometry, surface area, and field emission scanning electron microscopy/transmission electron microscopy image analyses: examples from the Barnett, Woodford, Haynesville, Marcellus, and Doig units[J]. AAPG Bull 96:1099–1119. https://doi.org/10.1306/10171111052

Chareonsuppanimit P, Mohammad SA, Robinson RL Jr et al (2012) High-pressure adsorption of gases on shales: measurements and modeling[J]. Int J Coal Geol 95:34–46. https://doi.org/10.1016/j.coal.2012.02.005

Chen B, Evans JRG, Greenwell HC et al (2008) A critical appraisal of polymer–clay nanocomposites[J]. Chem Soc Rev 37:568–594. https://doi.org/10.1039/B702653F

Chen T, Feng X, Pan Z (2018) Experimental study on kinetic swelling of organic-rich shale in CO_2, CH_4 and N_2[J]. J Nat Gas Sci Eng 55:406–417. https://doi.org/10.1016/j.jngse.2018.04.027

Clarkson CR, Solano NR, Bustin RM et al (2013) Pore structure characterization of North American shale gas reservoirs using USANS/SANS, gas adsorption, and mercury intrusion[J]. Fuel 103:606–616. https://doi.org/10.1016/j.fuel.2012.06.119

Cluff MA, Hartsock A, MacRae JD et al (2014) Temporal changes in microbial ecology and geochemistry in produced water from hydraulically fractured Marcellus shale gas wells [J]. Environ Sci Technol 48:6508–6517. https://doi.org/10.1021/es501173p

Cui G, Ren S, Rui Z et al (2018) The influence of complicated fluid-rock interactions on the geothermal exploitation in the CO_2, plume geothermal system[J]. Appl Energy 227:49–63. https://doi.org/10.1016/j.apenergy.2017.10.114

Deng H, Fitts JP, Crandall D et al (2015) Alterations of fractures in carbonate rocks by CO_2-acidified brines[J]. Environ Sci Technol 49:10226–10234. https://doi.org/10.1021/acs.est.5b01980

Duan S, Gu M, Du X et al (2016) Adsorption equilibrium of CO_2 and CH_4, and their mixture on Sichuan Basin shale[J]. Energy Fuel 30:2248–2256. https://doi.org/10.1021/acs.energyfuels.5b02088

Dustin M, Bargar J, Jew A et al (2018) Shale kerogen: hydraulic fracturing fluid interactions and contaminant release[J]. Energy Fuel 32:8966–8977. https://doi.org/10.1021/acs.energyfuels.8b01037

Feng G, Kang Y, Sun Z et al (2019) Effects of supercritical CO_2 adsorption on the mechanical characteristics and failure mechanisms of shale[J]. Fuel 173:870–882. https://doi.org/10.1016/j.energy.2019.02.069

Giesting P, Guggenheim S, Koster van Groos AF et al (2012) X-ray diffraction study of K- and Ca-exchanged montmorillonites in CO_2 atmospheres[J]. Environ Sci Technol 46:5623–5630. https://doi.org/10.1021/es3005865

Glikson M, Chappell BW, Freeman RS et al (1985) Trace elements in oil Shales, their source and organic association with particular reference to Australian deposits[J]. Chem Geol 53:155–174. https://doi.org/10.1016/0009-2541(85)90028-2

Goodman A, Sanguinito S, Tkach M et al (2019) Investigating the role of water on CO_2-Utica shale interactions for carbon storage and shale gas extraction activities – evidence for pore scale alterations[J]. Fuel 242:744–755. https://doi.org/10.1016/j.fuel.2019.01.091

Griffith CA, Dzombak DA, Lowry GV (2011) Physical and chemical characteristics of potential seal strata in regions considered for demonstrating geological saline CO_2 sequestration [J]. Environ Earth Sci 64:925–948. https://doi.org/10.1007/s12665-011-0911-5

He Y, Cheng J, Dou X et al (2017) Research on shale gas transportation and apparent permeability in nanopores[J]. J Nat Gas Sci Eng 38:450–457. https://doi.org/10.1016/j.jngse.2016.12.032

Heller R, Zoback M (2014) Adsorption of methane and carbon dioxide on gas shale and pure mineral samples[J]. J Unconv Oil Gas Res 8:14–24. https://doi.org/10.1016/j.juogr.2014.06.001

Hui D, Pan Y, Luo P et al (2019) Effect of supercritical CO_2 exposure on the high-pressure CO_2 adsorption performance of shales[J]. Fuel 247:57–66. https://doi.org/10.1016/j.fuel.2019.03.013

Huo P, Zhang D, Yang Z et al (2017) CO_2 geological sequestration: displacement behavior of shale gas methane by carbon dioxide injection[J]. Int J Greenh Gas Control 66:48–59. https://doi.org/10.1016/j.ijggc.2017.09.001

Ilgen A, Aman M, Espinoza D et al (2018) Shale-brine-CO_2 interactions and the long-term stability of carbonate-rich shale caprock[J]. Int J Greenh Gas Control 78:244–253. https://doi.org/10.1016/j.ijggc.2018.07.002

International Energy Agency, https://www.iea.org/topics/carbon-capture-and-storage/

Jarboe PJ, Candela PA, Zhu W et al (2015) Extraction of hydrocarbons from high maturity Marcellus shale using supercritical carbon dioxide[J]. Energy Fuel 29:7897–7909. https://doi.org/10.1021/acs.energyfuels.5b02059

Jean JS, Wang CL, Hsiang HI et al (2015) Experimental investigation of trace element dissolution in formation water in the presence of supercritical CO_2 fluid for a potential geological storage

site of CO_2 in Taiwan[J]. J Nat Gas Sci Eng 23:304–314. https://doi.org/10.1016/j.jngse.2015.02.006
Jiang F, Chen D, Chen J et al (2016a) Fractal analysis of shale pore structure of continental gas shale reservoir in the Ordos Basin, NW China[J]. Energy Fuel 30:4676–4689. https://doi.org/10.1021/acs.energyfuels.6b00574
Jiang Y, Luo Y, Lu Y et al (2016b) Effects of supercritical CO_2 treatment time, pressure, and temperature on microstructure of shale[J]. Energy 97:173–181. https://doi.org/10.1016/j.energy.2015.12.124
Jin L, Hawthorne S, Sorensen J et al (2017) Advancing CO_2, enhanced oil recovery and storage in unconventional oil play—experimental studies on Bakken shales[J]. Appl Energy 208:171–183. https://doi.org/10.1016/j.apenergy.2017.10.054
Jones TA, Detwiler RL (2016) Fracture sealing by mineral precipitation: the role of small-scale mineral heterogeneity[J]. Geophys Res Lett 43:7564–7571. https://doi.org/10.1002/2016GL069598
Kolak JJ, Burruss RC (2006) Geochemical investigation of the potential for mobilizing non-methane hydrocarbons during carbon dioxide storage in deep coal beds[J]. Energy Fuel 20:566–574. https://doi.org/10.1021/ef050040u
Lahann R, Mastalerz M, Rupp JA et al (2013) Influence of CO_2 on New Albany shale composition and pore structure[J]. Int J Coal Geol 108:2–9. https://doi.org/10.1016/j.coal.2011.05.004
Li Y, Li X, Wang Y et al (2015) Effects of composition and pore structure on the reservoir gas capacity of carboniferous shale from Qaidam Basin, China[J]. Mar Pet Geol 62:44–57. https://doi.org/10.1016/j.marpetgeo.2015.01.011
Liang C, Jiang Z, Zhang C et al (2014) The shale characteristics and shale gas exploration prospects of the lower Silurian Longmaxi shale, Sichuan Basin, South China[J]. J Nat Gas Sci Eng 21:636–648. https://doi.org/10.1016/j.jngse.2014.09.034
Liang M, Wang Z, Gao L et al (2017) Evolution of pore structure in gas shale related to structural deformation[J]. Fuel 197:310–319. https://doi.org/10.1016/j.fuel.2017.02.035
Liu F, Lu P, Griffith C et al (2012) CO_2–brine–caprock interaction: reactivity experiments on Eau Claire shale and a review of relevant literature[J]. Int J Greenh Gas Control 7:153–167. https://doi.org/10.1016/j.ijggc.2012.01.012
Liu D, Li Y, Agarwal RK (2016) Numerical simulation of long-term storage of CO_2 in Yanchang shale reservoir of the Ordos basin in China[J]. Chem Geol 440:288–305. https://doi.org/10.1016/j.chemgeo.2016.08.002
Liu Q, Tao L, Zhu H (2017) Macroscale mechanical and microscale structural changes in Chinese Wufeng shale with supercritical carbon dioxide fracturing[J]. SPE J 23:691–703. https://doi.org/10.2118/181369-pa
Lu Y, Ao X, Tang J et al (2016) Swelling of shale in supercritical carbon dioxide[J]. J Nat Gas Sci Eng 30:268–275. https://doi.org/10.1016/j.jngse.2016.02.011
Lu Y, Chen X, Tang J (2019) Relationship between pore structure and mechanical properties of shale on supercritical carbon dioxide saturation[J]. Energy 172:270–285. https://doi.org/10.1016/j.energy.2019.01.063
Luo X, Ren X, Wang S (2019) Supercritical CO_2-water-shale interactions and their effects on element mobilization and shale pore structure during stimulation[J]. Int J Coal Geol 202:109–127. https://doi.org/10.1016/j.coal.2018.12.007
Lyu Q, Ranjith PG, Long X et al (2016) Experimental investigation of mechanical properties of black shales after CO_2-water-rock interaction[J]. Materials (Basel) 9:663–678. https://doi.org/10.3390/ma9080663
Lyu Q, Long X, Ranjith PG et al (2018) Experimental investigation on the mechanical properties of a low-clay shale with different adsorption times in sub/super-critical CO_2[J]. Energy 147:1288–1298. https://doi.org/10.1016/j.energy.2018.01.084
Marcon V, Kaszuba JP (2015) Carbon dioxide–brine–rock interactions in a carbonate reservoir capped by shale: experimental insights regarding the evolution of trace metals[J]. Geochim Cosmochim Acta 168:22–42. https://doi.org/10.1016/j.gca.2015.06.037

Mokhtari M, Tutuncu AN (2015) Characterization of anisotropy in the permeability of organic-rich shales[J]. J Petrol Sci Eng 133:496–506. https://doi.org/10.1016/j.petrol.2015.05.024

Naik SN, Lentz H, Maheshawari RC (1989) Extraction of perfumes and flavors from plant materials with liquid carbon dioxide under liquid–vapor equilibrium conditions[J]. Fluid Phase Equilib 49:115–126. https://doi.org/10.1016/0378-3812(89)80009-3

Nuttall BC, Eble CF, Drahovzal JA et al (2005) Analysis of Devonian black shales in Kentucky for potential carbon dioxide sequestration and enhanced natural gas production. Report Kentucky Geological Survey/University of Kentucky (DE-FC26-02NT41442). https://doi.org/10.1016/B978-008044704-9/50306-2

Okamoto I, Li X, Ohsumi T (2005) Effect of supercritical CO_2 as the organic solvent on cap rock sealing performance for underground storage[J]. Energy 30:2344–2351. https://doi.org/10.1016/j.energy.2003.10.025

Pan Y, Hui D, Luo P et al (2018a) Influences of subcritical and supercritical CO_2 treatment on the pore structure characteristics of marine and terrestrial shales[J]. J CO2 Util 28:152–167. https://doi.org/10.1016/j.jcou.2018.09.016

Pan Y, Dong H, Luo P et al (2018b) Experimental investigation of the geochemical interactions between supercritical CO_2 and shale: implications for CO_2 storage in gas-bearing shale formations[J]. Energy Fuel 32:1963–1978. https://doi.org/10.1021/acs.energyfuels.7b03074

Phan TT, Capo RC, Stewart BW et al (2015) Trace metal distribution and mobility in drill cuttings and produced waters from Marcellus shale gas extraction: uranium, arsenic, barium[J]. Appl Geochem 60:89–103. https://doi.org/10.1016/j.apgeochem.2015.01.013

Rathnaweera TD, Ranjith PG, Perera MSA et al (2016) Experimental investigation of geochemical and mineralogical effects of CO_2 sequestration on flow characteristics of reservoir rock in deep saline aquifers[J]. Sci Rep 6:19362. https://doi.org/10.1038/srep19362

Rezaee R, Saeedi A, Iglauer S et al (2017) Shale alteration after exposure to supercritical CO_2[J]. Int J Greenh Gas Control 62:91–99. https://doi.org/10.1016/j.ijggc.2017.04.004

Rillard J, Gombert P, Toulhoat P et al (2014) Geochemical assessment of CO_2 perturbation in a shallow aquifer evaluated by a push–pull field experiment[J]. Int J Greenh Gas Control 21:23–32. https://doi.org/10.1016/j.ijggc.2013.11.019

Ripley EM, Shaffer NR, Gilstrap MS (1990) Distribution and geochemical characteristics of metal enrichment in the New Albany shale (Devonian-Mississippian), Indiana[J]. Econ Geol 85:1790–1807. https://doi.org/10.2113/gsecongeo.85.8.1790

Romanak KD, Smyth RC, Yang C et al (2012) Sensitivity of groundwater systems to CO_2: application of a site-specific analysis of carbonate monitoring parameters at the SACROC CO_2-enhanced oil field[J]. Int J Greenh Gas Control 6:142–152. https://doi.org/10.1016/j.ijggc.2011.10.011

Ross DJK, Bustin RM (2009) The importance of shale composition and pore structure upon gas storage potential of shale gas reservoirs[J]. Mar Pet Geol 26:916–927. https://doi.org/10.1016/j.marpetgeo.2008.06.004

Sanguinito S, Goodman A, Tkach M et al (2018) Quantifying dry supercritical CO_2-induced changes of the Utica shale[J]. Fuel 226:54–64. https://doi.org/10.1016/j.fuel.2018.03.156

Schaef H, Loring J, Glezakou V et al (2015) Competitive sorption of CO_2 and H_2O in 2:1 layer phyllosilicates[J]. Geochim Cosmochim Acta 161:248–257. https://doi.org/10.1016/j.gca.2015.03.027

Scherf AK, Zetzl C, Smirnova I et al (2011) Mobilisation of organic compounds from reservoir rocks through the injection of CO_2-comparison of baseline characterization and laboratory experiments[J]. Energy Procedia 4:4524–4531. https://doi.org/10.1016/j.egypro.2011.02.409

Shen Z, Sheng JJ (2017) Experimental study of permeability reduction and pore size distribution change due to asphaltene deposition during CO_2 huff and puff injection in Eagle Ford shale [J]. Asia Pac J Chem Eng 12:381–380. https://doi.org/10.1002/apj.2080

Shen Z, Sheng JJ (2018) Experimental and numerical study of permeability reduction caused by asphaltene precipitation and deposition during CO_2 huff and puff injection in Eagle Ford shale [J]. Fuel 211:432–445. https://doi.org/10.1016/j.fuel.2017.09.047

Shukla R, Ranjith P, Haque A et al (2010) A review of studies on CO_2 sequestration and caprock integrity[J]. Fuel 89:2651–2664. https://doi.org/10.1016/j.fuel.2010.05.012

Song W, Yao J, Jingsheng M et al (2018) Pore-scale numerical investigation into the impacts of the spatial and pore-size distributions of organic matter on shale gas flow and their implications on multiscale characterisation[J]. Fuel 216:707–721. https://doi.org/10.1016/j.fuel.2017.11.114

Soong Y, Crandall D, Howard BH et al (2017) Permeability and mineral composition evolution of primary seal and reservoir rocks in geologic carbon storage conditions[J]. Environ Eng Sci 35:391–400. https://doi.org/10.1089/ees.2017.0197

Span R, Wagner W (1996) A new equation of state for carbon dioxide covering the fluid region from the triple-point temperature to 1100 K at pressures up to 800 MPa[J]. J Phys Chem Ref Data Monogr 25:1509. https://doi.org/10.1063/1.555991

Strong L, Gould T, Kasinkas L et al (2013) Biodegradation in waters from hydraulic fracturing: chemistry, microbiology, and engineering[J]. J Environ Eng 140(5):B4013001-1-6. https://doi.org/10.1061/(asce)ee.1943-7870.0000792

Tan J, Weniger P, Krooss B et al (2014) Shale gas potential of the major marine shale formations in the upper Yangtze platform, South China, part II: methane sorption capacity[J]. Fuel 129:204–218. https://doi.org/10.1016/j.fuel.2014.03.064

Thomas MM, Clouse JA (1990) Primary migration by diffusion through kerogen: II. Hydrocarbon diffusivities in kerogen[J]. Geochim Cosmochim Acta 54:2781–2792. https://doi.org/10.1016/0016-7037(90)90012-A

Tian H, Pan L, Zhang T et al (2015) Pore characterization of organic-rich lower Cambrian shales in Qiannan depression of Guizhou Province, southwestern China[J]. Mar Pet Geol 62:28–43. https://doi.org/10.1016/j.marpetgeo.2015.01.004

Viswanathan H, Dai Z, Lopano C et al (2012) Developing a robust geochemical and reactive transport model to evaluate possible sources of arsenic at the CO_2 sequestration natural analog site in Chimayo, New Mexico[J]. Int J Greenh Gas Control 10:199–214. https://doi.org/10.1016/j.ijggc.2012.06.007

Wu T, Xue Q, Li X et al (2015) Extraction of kerogen from oil shale with supercritical carbon dioxide: molecular dynamics simulations[J]. J Supercrit Fluids 107:499–506. https://doi.org/10.1016/j.supflu.2015.07.005

Xiong J, Liu X, Liang L et al (2017) Adsorption of methane in organic-rich shale nanopores: an experimental and molecular simulation study[J]. Fuel 200:299–315. https://doi.org/10.1016/j.fuel.2017.03.083

Yang F, Ning Z, Zhang R et al (2015) Investigations on the methane sorption capacity of marine shales from Sichuan Basin, China[J]. Int J Coal Geol 146:104–117. https://doi.org/10.1016/j.coal.2015.05.009

Yang Z, Dong Z, Wang L et al (2018) Experimental study on selective adsorption/desorption of CO_2 and CH_4 behaviors on shale under a high-pressure condition[J]. Energy Fuel 32:9255–9262. https://doi.org/10.1021/acs.energyfuels.8b02068

Yin H, Zhou J, Jiang Y et al (2016) Physical and structural changes in shale associated with supercritical CO_2 exposure[J]. Fuel 184:289–303. https://doi.org/10.1016/j.fuel.2016.07.028

Yin H, Zhou J, Xian X et al (2017) Experimental study of the effects of sub- and super-critical CO_2 saturation on the mechanical characteristics of organic-rich shales[J]. Energy 132:84–95. https://doi.org/10.1016/j.energy.2017.05.064

Yu Z, Liu L, Yang S et al (2012) An experimental study of CO_2–brine–rock interaction at in situ pressure–temperature reservoir conditions[J]. Chem Geol 326–327:88–101. https://doi.org/10.1016/j.chemgeo.2012.07.030

Zhang K, Cheng Y, Li W et al (2017a) Influence of supercritical CO_2, on pore structure and functional groups of coal: implications for CO_2, sequestration[J]. J Nat Gas Sci Eng 40:288–298. https://doi.org/10.1016/j.jngse.2017.02.031

Zhang S, Xian X, Zhou J et al (2017b) Mechanical behaviour of Longmaxi black shale saturated with different fluids: an experimental study[J]. RSC Adv 7:42946–42955. https://doi.org/10.1039/c7ra07179e

Zhao T, Li X, Zhao H et al (2017) Molecular simulation of adsorption and thermodynamic properties on type II kerogen: influence of maturity and moisture content[J]. Fuel 190:198–207. https://doi.org/10.1016/j.fuel.2016.11.027

Zhou L, Bai SP, Su W (2003) Comparative study of the excess versus absolute adsorption of CO_2 on super-activated carbon for the near-critical region[J]. Langmuir 19:97–100. https://doi.org/10.1021/la020682z

Zhou J, Xie S, Jiang Y et al (2018) Influence of supercritical CO_2 exposure on CH_4 and CO_2 adsorption behaviors of shale: implications for CO_2 sequestration[J]. Energy Fuel 32:6073–6089. https://doi.org/10.1021/acs.energyfuels.8b00551

Ziemkiewicz PF (2013) Characterization of liquid waste streams from shale gas development [J]. AGH Drill Oil Gas 30:297–309. https://doi.org/10.7494/drill.2013.30.1.297

Zou Y, Li S, Ma X et al (2018) Effects of CO_2–brine–rock interaction on porosity/permeability and mechanical properties during supercritical-CO_2 fracturing in shale reservoirs[J]. J Nat Gas Sci Eng 49:157–168. https://doi.org/10.1016/j.jngse.2017.11.004

Chapter 7
Plantation Methods and Restoration Techniques for Enhanced Blue Carbon Sequestration by Mangroves

Abhiroop Chowdhury, Aliya Naz, and Santanu Bhattacharyya

Abstract Global warming and associated climate change is one of the focal environmental problems of this decade. So, Reducing Emissions from Deforestation and Forest Degradation (REDD) initiatives focuses on plantation programs in tropical/sub-tropical areas of the globe. Mangroves are group of plants with adaptations to survive in hyper-saline environment along the tropical/sub-tropical coastlines. As mangrove soil is anoxic with high salt content, C-sequestration potential of mangrove ecosystem is greater than any other terrestrial biomes. Litter fall and slow decomposition rate results in enriched organic carbon, organic matter, humic acid and fulvic acid in soil. It is reported that the litter alone contribute 2.07 Teragram of organic carbon enhancement at Sundarnbans mangrove ecosystem which is the world's largest contiguous mangrove forest and recently designated as a Ramsar wetland with site number 2370. Recent researches at Indian Sundarbans showed that the plantation of mangrove plants can increase the organic carbon density from 54–69 ton per ha to 83–132 ton per ha in a period of 5 years whereas the C-sequestration in mangrove sediments of Vietnam increased by 9.7% due to plantation activity with a high organic carbon density of 71–82 ton per ha.

Mangroves plantations shows low survival rate due to their sensitivity of environmental fluctuations. Establishing nursery with proper water drainage system, to raise propagules to approximately a height of 30–45 cm is advised before transplantation in plantation areas. *Rhizophora mucronata* a true mangrove of Rhizophoraceae family having stilt root adaptation, shows better survival if planted through trench and drainage method. *Avicennia marina* of Avicenniaceae family, can grow well with distribution of seedling directly after clearing a mudflat colonized by *Porteresia coarctata*, a Poaceae family pioneer species at Sundarban ecosystem. Fish bone canal can be used to restore sparse/degraded mangrove patches

A. Chowdhury (✉)
Department of Chemistry, Manipal University Jaipur, Jaipur, Rajasthan, India
e-mail: abhiroop.chowdhury@jaipur.manipal.edu

A. Naz
Department of Earth Science, Indian Institute of Technology, Kanpur, Uttar Pradesh, India

S. Bhattacharyya
Tagore Society for Rural Development, Kolkata, India

Inamuddin et al. (eds.), *Sustainable Agriculture Reviews 37*, Sustainable Agriculture Reviews 37, https://doi.org/10.1007/978-3-030-29298-0_7

by improving tidal water flow. Low survival rate of restored mangroves can be attributed to local factors as well as uncontrolled livestock grazing, growth of algal species like *Ulva* sp. and *Enteromorpha* sp. on young plants, infestation of crabs or poor water drainage in the sites. Hence post plantation care is required and participation of local community through group based organizations through mediation of local non-governmental organizations (NGO's) is effective in ensuring better survival of restored mangroves. Awareness campaigns to motivate and mobilize local populace prior to any plantation programs proved to be successful endeavor in respect of mangrove plantations.

Keywords Blue carbon · Mangroves · Ramsar site · Participatory forest management · In-situ conservation · Plantation methods · Post plantation care · Organic carbon · Soil carbon · Carbon sequestration · Community based organizations · Nursery · Carbon cycle · Non-governmental organization

7.1 Introduction

World is in verge of an environmental crisis while accommodating development and industries in cost of natural and natural resources. One of the foremost issue that is plaguing the world is the global warming and allied predicaments like increase of sea level, changes in weather pattern, increased instances of cyclones and natural disasters. In this context the foremost contributor to global warming is carbon di oxide (CO_2). Hence, there is a global impetus in sequestering CO_2 through different methods.

Development, population explosion and industrialization have resulted in loss of forested land globally and this process has accelerated in recent decades (Pendleton et al. 2012). These forests are a living repository of sequestered carbon and hence deforestation processes not only destroy the sequestered carbon but also results in reduction of carbon sink. Reasons for this unsustainable habitat conversion to facilitate human development vary globally. Aquaculture, agriculture, over-exploitation of forest, industrialization, construction of dams, dredging, and urban development are the principal anthropogenic activities that can be linked with habitat conversion (Pendleton et al. 2012; Valiela et al. 2001; Short and Wyllie-Echeverria 1996). There is a loss of 8000 km of forest land per year in the last century due to deforestation or land conversion due to anthropocentric purposes (Pendleton et al. 2012). This accounts for 0.5–3% loss of forest resource annually and results in 8–20% of global C-emission (Pendleton et al. 2012; Donato et al. 2011; Waycott et al. 2009; Van der Werf et al. 2009; Giri et al. 2008; IPCC 2007; Costanza et al. 1997).

Coastal regions are one of the most populated regions housing about 33% of human population in an area comprising of 4% of total land area of earth (Chowdhury et al. 2016a; Nicholls et al. 2007; Ericson et al. 2006). Blue carbon are carbon sequestered by oceanic and estuarine ecosytems. As this carbon sequestration is contributed by marine and estuarine ecosystems, it is termed as blue carbon after the color of the oceans. In case of coastal- estuarine habitats, it is generally referred to carbon trapped by seagrass meadows growing near shores and mangroves proliferating on tropical/subtropical mudflats. Mangrove forest (mangals) are plant community, with adaptations to survive in hypersaline conditions of coastal and estuarine regions in tropics and sub-tropics, and is responsible for ecological services like, filtering pollutants from soil-water continuum, saving the coasts from natural disasters like tsunamis or cyclones, reducing coastal erosion by tidal current and enhanced carbon sequestration in terms of biomass production as well as soil carbon trapping (Chowdhury and Maiti 2016a, b, c; Duarte et al. 2013; Donato et al. 2011; Cochard et al. 2008). Donato et al. 2011, indicated that estuarine regions holds 1074 MgC/ha and oceanic sites contained 990 MgC/ha. Mangroves stores highest C in their biomass and soil accounting to 1023 MgC/ha. This C- sequestration potential is highest if compared with mean C storage of all the world's other terrestrial ecosystems (Donato et al. 2011). In-spite of these ecological services mangroves are disappearing at an alarming rate and an estimate suggests that 30–40 percentage of salt marshes, sea grasses and nearly all of mangroves could disappear within a time span of 100 years (Pendleton et al. 2012; Duke et al. 2007; IPCC 2007).

Reducing greenhouse gasses emission and its sequestration is the focal point of the environmental policy debate in the last decade. In order to minimize unsustainable over utilization of natural resources, Reducing Emissions from Deforestation and Forest Degradation (REDD) initiative is gaining popularity in recent years. It operates on the concept that by reducing deforestation, the sequestered carbon is getting saved which would otherwise contribute to CO_2 load in the atmosphere. REDD initiatives is supposed to give developing countries opportunities to financially benefit from conserving their forest resources through direct payment or in-directly from marketing carbon stock in conserved plants (Turner et al. 2009). Hence, REDD initiates can slow down the accelerated rate of deforestation in coastal habitats and result in halting the process of climate change. Policy level changes have proved to be a successful tool in managing trans-boundary and global environmental issues. Montreal Protocol to restrict use and production of substances having deleterious effect on the Ozone Layer is effective from 26 August 1989, has proved to be a great success in combating ozone hole depletion issue. Recent researches highlights recovery of Ozone layer due to regulations restricting use of chlorofluorocarbons (CFCs), the main ozone depleting agent (Petrescu et al. 2018). So, REDD initiates can also transpires in a success story by restricting deforestation in vulnerable tropical, sub-tropical habitats.

Carbon sequestration is generally analyzed through above ground and below ground biomass of terrestrial ecosystems. But recent researches has highlighted that 'carbon' sequestered in anoxic estuarine, coastal or sea sediments takes a substantial part in global carbon budget and such soil/sediment carbon is termed as 'blue carbon. Alternatively, 'blue carbon' can be referred as carbon sequestered by living organisms and stored in sediments of coastal wetlands like mangroves, saline marshes and sea-grasses (Chowdhury et al. 2018; Chen et al. 2012; Nellemann et al. 2009). Sea-grass ecosystem and mangroves occupies only a small part of global coastline, but accounts for about 47% of total carbon burial (Duarte et al. 2013, 2008). For vegetated coastal lands, primary production exceeds respiration, hence, such floral assemblages can generates excess organic carbon acting as CO_2 sinks (Duarte et al. 2008). Burial of carbon particles flowing in through marine/tidal waters also adds up to the 'Blue carbon' sequestration budget in these coastal wetlands. Hence, coastal saline habitats have potential to sequester enormous amount of CO_2 which is almost 30–50 times that of terrestrial ecosystems (Chowdhury et al. 2018; Donato et al. 2011).

There is an exponential rise in plantation activities along the tropical coasts to restore/rehabilitate endangered mangroves as a consequence of REDD initiatives. But as these coastlines are mostly inhabited by socio-economically marginalized, disaster vulnerable population plantation works involving local community through Community Based Organizations (CBO's) proved to be a successful method in mangrove plantation works and these plantations not only act as a living carbon stock but also instrumental in exponential increase in 'blue-carbon' sequestration and also gives an economic benefit to native population. The benefit of blue carbon sequestration through mangrove plantation involving local population is discussed in this work.

7.2 Blue Carbon Sequestration

Coastal vegetation like mangroves, salt marshes and sea grasses that exist in the twilight zone between oceanic and terrestrial ecosystems takes a pivotal role in global carbon sequestration trapping unwanted CO_2 from environment, hence indirectly moderating the temperature change vagaries (Mcleod et al. 2011; Duarte et al. 2013; Nellemann et al. 2009; Duarte et al. 2008; Chmura et al. 2003). Alongi 2012, have defined 'Mangroves' as ecotone having attributes of both marine and terrestrial biomes having morphological, physiological and anatomical adaptations like viviparous germination, stilt root systems, salt tolerance, presence of salt glands, pneumatophores to survive in asphyxiated, saline environment. These plants sequester C within their underlying sediments along with above-ground (Stem, leaves, floral parts), below-ground (root) and dead tissues (litter, dead wood).

Living biomass can only store blue carbon for a short time span whereas sediment can act as C- sink for a larger time span (Mcleod et al. 2011; Duarte et al. 2005; Alongi 2012). Mean above-ground C pools in a mangrove ecosystem is around 159 MgC/ha, but below-ground biomass in soil accounts for 71–98% of total C storage in estuarine zones (Donato et al. 2011). Works of Donato et al. 2011 also shows that below-ground C storage was positively correlated to above-ground storage. Hence, plantation activities could not only result in increase above ground C- stock but also increases of bellow ground C- sequestration. Recent years have seen an increase of plantation of mangroves at the coastal fringes throughout the tropical-sub tropical region because of the ecosystem services rendered by the mangroves in term of saving the vegetated coasts from natural disasters like cyclones, typhoons, principal breeding ground of several marine and estuarine fish species, arresting erosion due to tidal/riverine current and blue carbon sequestration

(Cochard et al. 2008; Chowdhury et al. 2016a, 2018). The carbon sequestration budget of different mangrove ecosystems throughout the globe is elucidated in Table 7.1.

7.2.1 Carbon Balance in a Mangrove Ecosystem

Above ground biomass and C- sequestration is calculated by Net Primary Production (NPP) estimation which ranges between 453.6 and 101,605 kg dry weight/ha/year. According to Alongi 2012, most carbon sequestration assessment methods tend to overestimate (by equating NPP rates) or underestimate (litter-fall) the total carbon assimilation by biomass. For mangroves, the mean rate of aboveground NPP is 10,069.75 kg DW/ha/year, for tropical forests while mean aboveground NPP is around 10795.5 kg DW/ha/year (Alongi 2012). The general assumption during carbon sequestration estimation is that most of the carbon sequestration is allocated to the above ground biomass. Studies highlight that mangroves sequester more carbon in their belowground biomass than the terrestrial trees (Alongi 2012; Donato et al. 2011; Lovelock 2008). Growth of the plants results in production of organic matter due to C- fixation via photosynthesis. Mangroves are evergreen plants that keep on adding litter load in the environment by which the organic matter/organic carbon concentration increases in the soil (Inoue 2019). But due to elevated pH, high salt content and O_2 deficiency in soil decomposition process is hindered that keeps the semi decomposed plant matter trapped in the soil. Mangrove plantation aid in increased sequestration of soil carbon via organic matter, humic acid (HA) and fulvic acid (FA) fractions within a time span of 5–6 years, as evident from a recent study by Chowdhury et al. 2018. The pathways for C- sequestration in mangrove are elucidated in Fig. 7.1.

7.3 Plantation Techniques for Mangroves

Mangrove shows a low survival after restoration measures. Hence, adoption of proper plantation methods along with post plantation care helps in maximizing the success rate of mangrove plantation/restoration efforts. The greatest challenge in reforestation of mangrove lands is the low survival of seedling/sapling as evident from current literatures on mangrove plantations (Chowdhury et al. 2018). Another attribute that plays a deciding role in plantation methods is the economic footprint of the particular method. Generally, more survival efficiency is required by implementation organizations with maximum carbon sequestration in cheapest cost which substantiates importance of plantation methods in mangrove reforestation. Different methods for mangrove plantation is given below.

Table 7.1 Studies on worldwide mangrove plantation and survey in the context of Carbon sequestration

Reference	Country	Sampling area	Area (ha)	Major plant Sp.	[a]Year of Plantation/[b]year of survey	Average Height (m)	Comments
Phan et al. (2019)	Vietnam	Six coastal provinces of Mekong delta in Vietnam	0.5–1.5	*Rhizophora apiculata*	[a]1988–2009	5.4–15.5	Sediment carbon sequestration in the range of 3.9–22.7 mg/ha/year
Kathiresan et al. (2018)	India	Northern bank of the Vellar estuary, located along the Bay of Bengal on the southeastern coast of the state of Tamil Nadu	15	*Avicennia marina and Rhizophora mucronata*	[b]2009–2010	0.3–1.3	Carbon sequestration about 6.12 ± 3.36 kg/tree/year for *Avicennia marina* and 12.05 ± 3.55 kg/tree/year for *Rhizophora mucronata*
Peng et al. (2016)	China	Yifeng Estuary of Shantou City	22.9	*Kandelia obovata,*	[a]2004	7.76	Total organic carbon (TOC) in sediments was 48.7 ± 10.1 (g/kg)
			33.4	*Aegiceras corniculatum*	[a]2008	4.82	TOC in sediments was 45.3 ± 11.3 (g/kg)
			15.2	*Bruguiera gymnorhiza*	[a]2010	3.43	TOC in sediments was 40.6 ± 8.49 (g/kg)
Chen et al. (2012)	China	Intertidal zones of Shenzhen Bay, Guangdong Province, China	NA	Native *S. caseolaris* and non-native *S. apetala*	[a]2005–2007	2.1–3.78	Organic carbon was in the range of 0.12 to 14.55%
Eid and Shaltout (2016)	Egypt	Red Sea Coast Ras Muhammed, and the southern section of the Gulf of Aqaba	52,500	*Avicennia marina* (black mangrove) and *Rhizophora mucronata* (red mangrove)	NA	1.5–3.5	Carbon sequestration estimated to be 6.1 g Carbon/m^2/year)

Nguyen et al. (2004)	Vietnam	Giaolac Commune, Namdinh Province located in the southeast section of the Red River Delta in Northern Vietnam	347, 140, 300,	*Kandelia candel (L.), Sonneratia caseolaris (L.) and Rhizophora stylosa*	[a]1992–1998	1.1–3.92	Sequester carbon upto 71 to 82 (ton Carbon per ha)
Ouyang et al. (2017)	Australia	Along Tallebudgera Creek, Southeast Queensland, Australia	NA	*Avicennia marina and Rhizophora stylosa*	[b]2016	NA	Carbon was estimated 2994 ± 186 g/m^2 in the sediment of *Avicennia marina* and 2383 ± 209 g/m^2 was estimated in *Rhizophora mucronata*
Kusumaningtyas et al. (2018)	Indonesia	Undisturbed forest in Berau East Kalimantan	NA	Area dominated by *Sonneratia alba*	[b]2013	NA	Total organic carbon in sediment was 722 ± 183 g/m^2/year
		Segara Anakan Lagoon (SAL) southern Central Java	9000	Area dominated by *Sonneratia caseolaris*	[b]2016	NA	Total organic carbon in sediment was 658 ± 311 g/m^2/year
Ray et al. (2011)	India	Sundarbans	420,000	*Avicennia marina, Avicennia alba, Avicennia officinalis, Ceriops sp., Excoecaria agallocha, Excoecaria agallocha, Agialitis rotundifolia*	[b]2009–2010	4.5–10.1	Total carbon storage in the sediment was estimated to be 5.49 Tg C with an addition rate of 2.07 Tg C in terms of litter fall.
Barreto et al. (2016)	Venezuela	Venezuelan Caribbean coast along the Cuare Inlet and Morrocoy National Park	NA	*Rhizophora and Avicennia*	[b]2005–2006	3–15.5	Average sediment organic matter was 0.75 ± 0.04 wt%

(continued)

Table 7.1 (continued)

Reference	Country	Sampling area	Area (ha)	Major plant Sp.	[a]Year of Plantation/[b]year of survey	Average Height (m)	Comments
Chowdhury et al. (2018)	India	Indian Sunderbans	102.4	*Avicennia marina, Bruguiera sexangula, Ceriops tagal, Rhizophora mucronata and Xylocarpus moluccensis*	[a]2012–2016	1.3–2.52	The organic carbon density in sediments was 54 and 69 t/ha in 2012; which increased to 83–132 t/ha over a period of 5 years.

[a]Year of plantation
[b]Year of the survey
NA not available

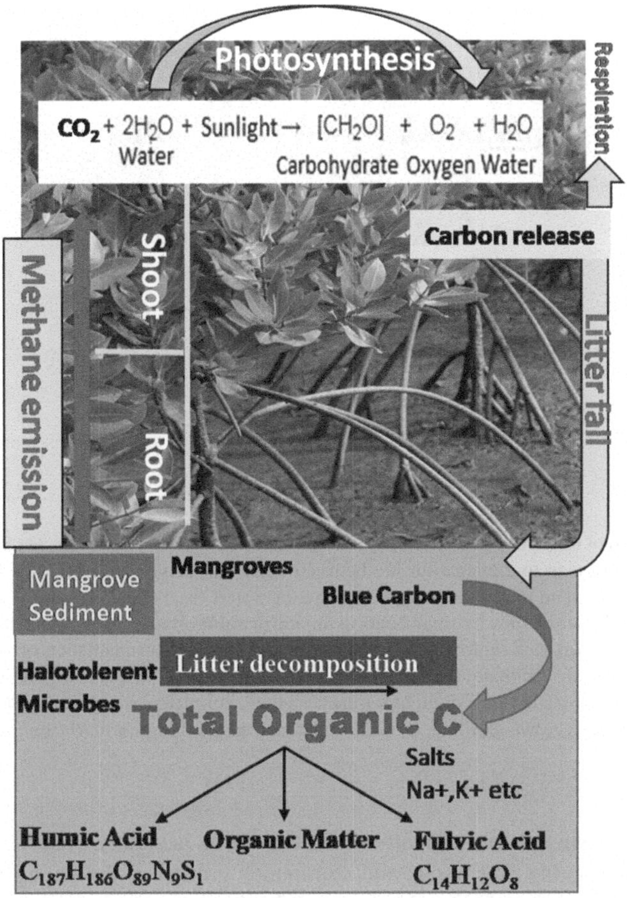

Fig. 7.1 CO_2 is sequestered by the evergreen mangrove plants through the process of photosynthesis. A part of the sequestered energy is lost due to the metabolic process of respiration and litter fall. Litter is decomposed through the mediation of halotolerent microbes in the mangrove soil in presence of high salinity (Sodium, Potassium, Magnesium, Calcium ions). This decomposition process takes longer time than terrestrial ecosystem due to the high salinity and alkaline environment. The semi decomposed plant materials exist in anoxic mangrove edaphic environment in form of Humic Acid, Fulvic acid or undecomposed Organic matter which constitute the Blue carbon pool in this ecosystem

7.3.1 Establishment of Mangrove Nursery

The most reliable method of plantation is transplantation of nursery raised seedling (Fig. 7.2). But this method is costly and labor intensive that reduces the popularity of this method in many of the plantation sites. Pre-requisite of this method is establishment of a nursery near to the plantation sites where the propagules can be raised, tended to a particular height before transplanting them to the plantation sites. Most of

Fig. 7.2 Different mangrove plantation methods practiced at Indian sundarbans; (**a**) Setting up of a mangrove nursery near to a plantation site between High Tide Level and Low Tide Level near at side of creek system so that it can get steady flow of water, (**b**) Transportation of nursery raised saplings to the plantation sites by small boats to minimize transportation stress to seedlings, (**c**) local population taking part in mangrove plantation by digging pits for transplantation of nursery grown seedlings/saplings to facilitate participatory forest conservation, (**d**) successful plantation of *Rhizophora mucronata* through drain and trench method, (**e**) mangrove plantation through transplantation of nursery grown seedling, (**f**) drain and trench earthwork prior to *Rhizophora mucronata* plantation

the true mangrove species viviparous germination is generally observed. Vivipary can be defines as the precocious and continuous growth of the young plant while connected to the maternal plant (Naskar et al. 2000). So a seedling can germinate and grow for a period of time before dispersal, like in *Rhizophora* sp. This adaptation of mangrove plants facilitate their survival in hyper saline stress and to maximize the survival of the seedling in a asphyxiated edaphic environment which is classified as a physiological dry soil because of inaccessibility of the water to plants due to elevated salt content. Nurseries should be developed in s bed of less than a thousand sapling in water proof bags/pots along the tidal water flow like creeks, channels, distributary networks so that the growing sapling can be regularly get flushing of water which is a major requirement for unhindered and proper growth (Fig. 7.2a).

Selection of nursery sites close to the plantation sites ensure a lower labor investment while transportation as well as less stress on young saplings (Fig. 7.2b). Precautions must be taken for drainage of inlet and outlet water in the nursery beds for proper drainage of water after the tidal water recedes. Water logging can result in death of the seedling/saplings. Propagules need to be collected from healthy tree during onset of monsoon/rainy season in tropics and they should be free from microbial/entomological infections. Generally yellowish green propagules should be checked for such infection before growing them in nurseries to ensure a better survival of the plants at later stages (Thivakaran et al. 2016). It is always advisable to collect mature seeds that has already dislodged itself from the parent plant as they shows proper sign of germination. In case of reserve forest areas like Indian Sundarbans only process of collecting seeds would be one that floats in the human habituated areas adjoining the forest areas during tides as forest mangroves would not be accessible to local populations because of conservation regulations. Indian Sundarbans has been designated as a Ramsar site (site number: 2370) on February, 2019 owing to its ecological importance and is already an United Nations Educational, Scientific and Cultural Organization (UNESCO) world heritage site as it is the only abode of mangrove inhabiting tigers (*Panthera tigris tigris*) and the largest contiguous mangrove forest of the world.

Polythene bags (5 × 8 inches) dimension is stuffed with clayey soil collected from the nearby mudflat before placing the seeds. Dimensions of nursery beds vary with different plantation zones. For example at riverine Indian Sundarbans it has a dimension of 10 m (length) × 1 m (breadth) × 20 cm (depth) whereas semi-arid mangrove formations at Kachchh, Gujarat (India) practice raised nursery beds of about 20 cm (Thivakaran et al. 2016; Chowdhury et al. 2018). Beds should be created in partial shaded conditions to ensure maximum survival of the saplings.

7.3.2 Transplantation of Nursery Grown Seedlings

Transplantation of nursery grown seedling is generally starts when the saplings attain 30–45 cm in height. In the plantation sites holes are dug of same size little more than the dimension of the bags. In the case of Indian Sundarbans pits of 1 ft × 1 ft × 1 ft pit is dug before transplantation (Thivakaran et al. 2016; Chowdhury et al. 2018). Plantation could be done through participation of local population (Fig. 7.2c). Utmost care need to be taken during the transplantation process to ensure the fragile root system would not be damaged during the plantation. Seedling need to be planted till the emergence point of the shoot system. Generally 5.5–6.0 ft gap is to be maintained between the pits, so that the plants would not be congested in a small area and get ample space to grow without competition (Chowdhury et al. 2018). If that size is maintained 3500 saplings can be planted in a 1 ha area. As per recent studies this method results in the maximum survival of the plants and is superior to other plantation methods though it has more expensive because of the expenditure on labor involvement and nursery maintenance.

7.3.3 Direct Seeding Method

Direct seeding method has a lesser economic liability to plantation program and has a better cost-budget output if cost of plantation and carbon sequestration rate is weighted together. This is evident from recent study of Chowdhury et al. 2018, on efficiency of mangrove plantation at the largest mangrove forest zone of the globe (Sundarbans, India-Bangladesh). In this method the mudflat is cleared of *Porteresia coarctata* (Family: Poaceae) plants and then propagules are directly distributed on the plantation site. *P. coarctata* is a halotolerant grass and pioneer species in the succession pattern at tropical mangrove ecosystems like Sundarbans and the extensive root matting of this species hinders the establishment of propagules of other mangrove species. Hence, it is necessary to clear this grass patch before the advent of plantation activities. Only *Avicennia marina* (Family: Avicenniaceae), shows a high survival in this method. *A. marina* propagules are easily available as it is the abundant mangrove in both the geographical locations of Indian Sundarbans, West Bengal and Kachchh coast, Gujarat. Local fishermen collect the floating propagules from the tidal waters, washed out of the reserve forest area and store them in gunny bags. One bag generally can hold approximately 2000 propagules (Chowdhury et al. 2018).

7.3.4 Drain and Trench Method

Mangrove plants have various adaptations to survive in a hyper-saline environment (Tomlinson 2016; Naskar et al. 2000; Chowdhury et al. 2016a, b). *Rhizophora mucronata* (Family: Rhizophoraceae) has a typical adaptation of supporting stilt roots coming out from the shoot system. Hence, in parts of Indian Sundarbans, drains and trenches are dug before transplantation of nursery raised seedling to give space for stilt roots to spread (Fig. 7.2d).Stilt roots are not damaged in the process as evident from a plantation activity at Indian Sundarbans (Fig. 7.2d). Such method can improve the survival rate of *R. mucronata*to 55% but has a larger economic liability on the plantation project as evident from the study of Chowdhury et al. 2018.

7.3.5 Fish Bone Canal System

Fish bone cannel method has been extensively used to restore degraded mangrove areas. The main canals are dug 45° to the main creek and side canals are generally dug 30° to the main canal (Thivakaran et al. 2016). The architecture of the canals mimics the structure of a fishbone. This facilitates the easy inflow and outflow of water (Ramasubramanian and Ravishankar 2004). Fish bone canal would produce better result if the drainage area has a trough shaped with sparse mangroves. The level of the drainage area should be below the Mean High Tide Level (MHTL), so that regular flushing of the water is possible through the channel system even to the feeder channel. This method is only possible when the periphery of the restoration area is at an elevated level, hence hindering the flow of tidal water into the inner parts which is vital for the proliferation and growth of mangroves. Coastal mudflats beyond the intertidal zone are not suitable for this method of restoration. Similarly, sandy shores are also not suitable for fish bone canal system as the sand structures would collapse with rain fall and sedimentation would clog the waterways. If the mudflats are on heavy sedimentation pressure the channel mouth can be shut due to sediment deposition. Local population and forest authorities are well versed with the land topography, who can aid in selection of sites for successful execution of fishbone canals.

7.4 Post Plantation Management

Post plantation stress should be minimized after the plantation of mangroves. As these halophytes are sensitive to abiotic and biotic stressors, utmost care should be taken for maintaining the plantations. In most of the cases, post plantation funding is limited but recent studies has shown that fund allocation to post plantation care improves the survival of the plants as well as ensures a healthy growth (Chowdhury et al. 2018).

Livestock grazing is one of the major problems for the survival of the mangrove plants in most of the mudflats. The solution would be extensive awareness campaigns amongst the local population before advent of plantation projects (Chowdhury et al. 2016b, 2018). Participatory methods where the local populations took part in the plantation through group based mobilization or as a labor, works wonderfully well in modifying the local populace's perspective on the plantation programs as is evident from such endeavors at Indian Sundarbans (Chowdhury et al. 2016b, 2018). Encrustation of algae and barnacles are a serious threat for the survival of the young transplanted sapling/seedlings. The algal species mostly *Ulva* sp., and *Enteromorpha* sp. is seen to grow on young plants along Kachchh coastline semi-arid marine communities of Gujarat, India (Thivakaran et al. 2018). Barnacles generally seen to settle on hard substratum, hence tree trunks are mostly affected by them. Leaves and exposed soft tissues are safe from such infestations. Algal infestations can be physically removed by hand but there are no such remedial measures for barnacle infestations (Thivakaran et al. 2016).

Crabs and mollusks plays an integral role in maintenance of a mangrove ecosystem as they cut down the detritus into finer particles. Sometimes crabs can also affect the root system of the plants. Works of Ong et al. 2010; Dahdouh-Guebas and Satyanarayana 2012, have discussed the crab (*Neosarmatium meinerti*) that predates voraciously on the mangrove propagules at restoration sites at Kenya. At restoration sites at Gulf of Kachchh (Gujarat, India) species of *Ocypode* sp. feeds on germinating propagules. Van Nedervelde et al. 2015, argued that crab density is proportional to the density of vegetation and crabs can negatively influence the survival, health of the young plants. But if mangrove ecosystem can be considered holistically, proliferation of crabs in a mangrove ecosystem is indicative of a healthy ecosystem for their pivotal role in energy cycling in mangrove ecosystem through detritous food chain, hence crabs cannot be treated as pest despite of their damage on plantation efforts (Kristensen 2008). Solution to this problem is to replace the damaged seedlings instead of eliminating the crabs in the plantation sites. Mangroves shows variations in structure and function with differences in abiotic factors like moisture content of soil, types of soil, nutrient concentrations and the same also influences the survival rate of plants after a restoration program (Chowdhury et al. 2018; Thivakaran et al. 2018).

7.5 Community Participation in Mangrove Plantation

Community can be motivated in plantation programs. Mostly the range of mangroves is in developing nations between the tropical and sub tropical zone in populated coasts. The population pressure, paucity of employment opportunities, financial marginality imbibes the local populations with apathy towards conservation of natural resources even if there is legislative framework for that (Chowdhury et al. 2016a). Many local practices are sometimes are against the conservation objectives. For example, tiger prawn (*Penaeus monodon*) seed collection one of a popular trade in waterways and creeks of sundarbans at Ganges-Brahmaputra river system. The practice is to drag a net in mud across the water flowing out after high tide against the water current to collect small tiger prawn larva in it, to be sold in aquaculture farms (Chowdhury et al. 2017). But due to its negative impact on mangrove ecosystem, this method of fishing was restricted through legislations by Indian government. But due to paucity of employment options and income sources local residents are still engaged in that trade with negative health as well as ecological impacts (Chowdhury et al. 2017). But awareness campaigns and popularization of green sustainable livelihood through group based organizations along with implementing mangrove restoration programs proved to be a successful method as evident from recent works of Chowdhury et al. 2016b, 2018. Local population can actively take part in plantation programs and can be instrumental in post plantation care. As mangroves are first natural shield of defense on advent of any natural disasters like cyclones, storm surges, people can be enlighten on the benefits of mangroves through awareness campaigns prior to plantation work (Chowdhury et al. 2016a, b). Planation programs involving local populations are an effective mangrove restoration process around the globe because they have prior traditional knowledge on the area, environmental issues and can actively aid in post plantation care (Kamali and Hashim 2011; Ren et al. 2011; Kairo et al. 2008; Gilman and Ellison 2007; Walters 2000; Ellison 2000).

7.6 Conclusion

Mangroves are one of the most important carbon sink as evident from previous researches existing in the borderland between land and sea. Mangrove plants show low responses to plantation efforts in degraded coastlines in term of survival rate and growth. But community based plantation endeavors can give a solution to these limitations. Several plantation techniques are adopted throughout globe for survival of mangrove plants with maximum money to carbon sequestration conversion rate, i.e. optimizing the economic efficiency of plantation methods in terms of carbon sequestration. Different mangrove plants show varied responses to plantation method. For example, *Rhizophora mucronata* have a better survival rate through drain and trench method whereas *Avicennia marina*, shows better survival in propagule dibbling method. Successful implementation of nursery near to plantation sites give a benefit to plantation endeavors, but nursery set up needs proper water flow and maintenance. Post plantation care is essential to ensure the maximum survival of the plants as they are affected by crab attacks, barnacle encrustation, algal clogging and livestock grazing. Community involvement through group based organizations and awareness campaigns prior to plantation programs proved to be successful in enlisting local support and also to maximize the survival of plants.

Acknowledgement We sincerely thank Mr. Martin Wolff and Soumyendra Roy for their support. We are thankful for the support provided by Tagore Society for Rural Development (TSRD), KKS (Karl Kübel Stiftung für kind und de Familie, Germany) and BMZ (Federal Ministry for Economic Cooperation and Development, Germany) from 2012 to 2017 through "Peoples Empowerment Towards Restoring Mangrove Vegetation and Resource Conservation" program implemented at Indian Sundarbans.

Conflict of Interest There is no conflict of interest to declare.

References

Alongi DM (2012) Carbon sequestration in mangrove forests. Carbon Manage 3:313–322. https://doi.org/10.4155/cmt.12.20

Barreto MB, Mónaco SL, Díaz R, Barreto-Pittol E, López L, Peralba MDCR (2016) Soil organic carbon of mangrove forests (*Rhizophora and Avicennia*) of the Venezuelan Caribbean coast. Org Geochem 100:51–61. https://doi.org/10.1016/j.orggeochem.2016.08.002

Chen L, Zeng X, Tam NF, Lu W, Luo Z, Du X, Wang J (2012) Comparing carbon sequestration and stand structure of monoculture and mixed mangrove plantations of *Sonneratia caseolaris* and *S. apetala* in Southern China. For Ecol Manag 284:222–229. https://doi.org/10.1016/j.foreco.2012.06.058

Chmura GL, Anisfeld SC, Cahoon DR, Lynch JC (2003) Global carbon sequestration in tidal, saline wetland soils. Global Biogeochem Cycles 17:1111. https://doi.org/10.1029/2002GB001917

Chowdhury A, Maiti SK (2016a) Assessing the ecological health risk in a conserved mangrove ecosystem due to heavy metal pollution: a case study from Sundarbans biosphere reserve, India. Hum Ecol Risk Assess Int J 22(7):1519–1541. https://doi.org/10.1080/10807039.2016.1190636

Chowdhury A, Maiti SK (2016b) Identifying the source and accessing the spatial variations, contamination status, conservation threats of heavy metal pollution in the river waters of Sunderban biosphere reserve, India. J Coast Conserv 20(3):257–269. https://doi.org/10.1007/s11852-016-0436-x

Chowdhury A, Maiti SK (2016c) Identification of metal tolerant plant species in mangrove ecosystem by using community study and multivariate analysis: a case study from Indian Sunderban. Environ Earth Sci 75(9):1–21. https://doi.org/10.1007/s12665-016-5391-1

Chowdhury A, Sanyal P, Maiti SK (2016a) Dynamics of mangrove diversity influenced by climate change and consequent accelerated sea level rise at Indian Sundarbans. Int J Glob Warm 9 (4):486–506. https://doi.org/10.1504/IJGW.2016.076333
Chowdhury A, Maiti SK, Bhattacharyya S (2016b) How to communicate climate change 'impact and solutions' to vulnerable population of Indian Sundarbans? From theory to practice. Springerplus 5:1219. https://doi.org/10.1186/s40064-016-2816-y
Chowdhury A, Naz A, Maiti SK (2017) Health risk assessment of 'tiger prawn seed'collectors exposed to heavy metal pollution in the conserved mangrove forest of Indian Sundarbans: a socio-environmental perspective. Hum Ecol Risk Assess Int J 23(2):203–224. https://doi.org/10.1080/10807039.2016.1238300
Chowdhury A, Naz A, Bhattacharyya S, Sanyal P (2018) Cost benefit analysis of 'blue carbon' sequestration by plantation of few key mangrove species at Sundarban biosphere reserve, India. Carbon Manage 9(6):575–586. https://doi.org/10.1080/17583004.2018.1518105
Cochard R, Ranamukhaarachchi SL, Shivakoti GP, Shipin OV, Edwards PJ, Seeland KT (2008) The 2004 tsunami in Aceh and Southern Thailand: a review on coastal ecosystems, wave hazards and vulnerability. Perspect Plant Ecol Evol Syst 10(1):3–40. https://doi.org/10.1016/j.ppees.2007.11.001
Costanza R, d'Arge R, de Groot R, Farber S, Grasso M et al (1997) The value of the world's ecosystem services and natural capital. Nature 387:253–260. https://doi.org/10.1016/S0921-8009(98)00020-2
Dahdouh-Guebas F, Satyanarayana B (2012) Proceedings of the international conference' meeting on Mangrove ecology, functioning and Management-MMM3', Galle, Sri Lanka, 2–6 July 2012. VLIZ Special Publication. Vlaams Instituut voor de Zee (VLIZ): Oostende. ISSN:1377-0950
Donato DC, Kauffman JB, Murdiyarso D, Kurnianto S, Stidham M, Kanninen M (2011) Mangroves among the most carbon-rich forests in the tropics. Nat Geosci 4(5):293–297. https://doi.org/10.1038/ngeo1123
Duarte CM, Middelburg JJ, Caraco N (2005) Major role of marine vegetation on the oceanic carbon cycle. Biogeosciences 2(1):1–8. https://doi.org/10.5194/bg-2-1-2005
Duarte CM, Dennison WC, Orth RJW, Carruthers TJB (2008) The Charisma of coastal ecosystems: addressing the imbalance. Estuar Coasts 31(2):233–238
Duarte CM, Losada IJ, Hendriks IE, Mazarrasa I, Marbà N (2013) The role of coastal plant communities for climate change mitigation and adaptation. Nat Clim Chang 3(11):961–968. https://doi.org/10.1038/nclimate1970
Duke NC, Meynecke JO, Dittmann S, Ellison AM, Anger K et al (2007) A world without mangroves? Science 317:41–42. https://doi.org/10.1126/science.317.5834.41b
Eid EM, Shaltout KH (2016) Distribution of soil organic carbon in the mangrove *Avicennia marina* (Forssk.) Vierh. along the Egyptian Red Sea Coast. Reg Stud Mar Sci 3:76–82. https://doi.org/10.1016/j.rsma.2015.05.006
Ellison AM (2000) Mangrove restoration: do we know enough? Restor Ecol 8(3):219–229. https://doi.org/10.1046/j.1526-100x.2000.80033.x
Ericson J, Vorosmarty C, Dingman S, Ward L, Meybeck M (2006) Effective Sea-level rise and deltas: causes of change and human dimension implications. Glob Planet Chang 50(1–2):63–82
Gilman E, Ellison J (2007) Efficacy of alternative low-cost approaches to mangrove restoration, American Samoa. Estuar Coasts 30(4):641–651. https://doi.org/10.1007/BF02841961
Giri C, Zhu Z, Tieszen LL, Singh A, Gillette S et al (2008) Mangrove forest distributions and dynamics (1975–2005) of the tsunami-affected region of Asia. J Biogeogr 35:519–528. https://doi.org/10.1111/j.1365-2699.2007.01806.x
Inoue T (2019) Carbon sequestration in mangroves. In: Blue carbon in shallow coastal ecosystems. Springer, Singapore, pp 73–99. https://doi.org/10.1007/978-981-13-1295-3-3
Intergovernmental Panel on Climate Change (IPCC) (2007) Climate change 2007: assessment report. IPCC, Valencia. IPCC in The Fourth Assessment Report Climate Change 2007 (eds) Pachauri RK and Reisinger A (IPCC, 2007)
Kairo JG, Lang'at JK, Dahdouh-Guebas F, Bosire J, Karachi M (2008) Structural development and productivity of replanted mangrove plantations in Kenya. For Ecol Manag 255(7):2670–2677. https://doi.org/10.1016/j.foreco.2008.01.031

Kamali B, Hashim R (2011) Mangrove restoration without planting. Ecol Eng 37(2):387–391. https://doi.org/10.1016/j.ecoleng.2010.11.025

Kathiresan K, Saravanakumar K, Asmathunisha N, Anburaj R, Gomathi V (2018) Biochemical markers for carbon sequestration in two mangrove species (Avicennia marina and Rhizophora mucronata). Beni-Suef Univ J Basic Appl Sci 7(4):733–739. https://doi.org/10.1016/j.bjbas.2018.10.003

Kristensen E (2008) Mangrove crabs as ecosystem engineers; with emphasis on sediment processes. J Sea Res 59(1–2):30–43. https://doi.org/10.1016/j.seares.2007.05.004

Kusumaningtyas MA, Hutahaean AA, Fischer HW, Pérez-Mayo M, Pittauer D, Jennerjahn TC (2018) Variability in the organic carbon stocks, sources, and accumulation rates of Indonesian mangrove ecosystems. Estuar Coast Shelf Sci. https://doi.org/10.1029/2012GB004375

Lovelock CE (2008) Soil respiration and belowground carbon allocation in mangrove forests. Ecosystems 11:342–354. https://doi.org/10.1007/s10021-008-9125-4

Mcleod E, Chmura GL, Bouillon S, Salm R, Björk M, Duarte CM et al (2011) A blueprint for blue carbon: toward an improved understanding of the role of vegetated coastal habitats in sequestering CO_2. Front Ecol Environ 9(10):552–560. https://doi.org/10.1890/110004

Naskar K, Sarkar D, Thakur GC, Sen N (2000) Identification and impact of the algal flora of the Indian Sundarbans. Sundarban Mangal:350–354

Nellemann C, Corcoran E, Duarte CM et al (eds) (2009) Blue carbon. A rapid response assessment. GRID-Arendal: United Nations Environment Programme. ISBN:978-82-7701-060-1

Nguyen HT, Yoneda R, Ninomiya I, Harada K, Van Dao T, Sy TM, Phan HN (2004) The effects of stand-age and inundation on carbon accumulation in mangrove plantation soil in Namdinh, Northern Vietnam. Tropics 14(1):21–37

Nicholls RJ, Wong PP, Burkett VR, Codignotto JO, Hay JE, McLean RF, Ragoonaden S, Woodroffe CD (2007) Coastal systems and low-lying areas. Climate change 2007: impacts, adaptation and vulnerability. In: Parry ML, Canziani OF, Palutikof JP, van der Linden PJ and Hanson CE (eds) Contribution of Working Group II to the Fourth Assessment Report of the Intergovernmental Panel on Climate Change, pp 315–356

Ong SP, Cheng S, Chong VC (2010) Pests of planted mangroves in peninsular Malaysia. Forest Research Institute Malaysia. ISBN:978-967-5221-40-8

Ouyang X, Lee SY, Connolly RM (2017) Structural equation modelling reveals factors regulating surface sediment organic carbon content and CO^2 efflux in a subtropical mangrove. Sci Total Environ 578:513–522. https://doi.org/10.1016/j.scitotenv.2016.10.218

Pendleton L, Donato DC, Murray BC, Crooks S, Jenkins WA, Sifleet S et al (2012) Estimating global "blue carbon" emissions from conversion and degradation of vegetated coastal ecosystems. PLoS One 7(9):e43542. https://doi.org/10.1371/journal.pone.0043542

Peng Y, Diao J, Zheng M, Guan D, Zhang R, Chen G, Lee SY (2016) Early growth adaptability of four mangrove species under the canopy of an introduced mangrove plantation: implications for restoration. For Ecol Manag 373:179–188. https://doi.org/10.1016/j.foreco.2016.04.044

Petrescu RV, Aversa R, Apicella A, Petrescu FI (2018) Nasa sees first in 2018 the direct proof of ozone hole recovery

Phan SM, Nguyen HTT, Nguyen TK, Lovelock C (2019) Modelling above ground biomass accumulation of mangrove plantations in Vietnam. For Ecol Manag 432:376–386. https://doi.org/10.1016/j.foreco.2018.09.028

Ramasubramanian R, Ravishankar T (2004) Mangrove forest restoration in Andhra Pradesh, India. MS Swaminathan Research Foundation, Chennai, MSSRF/MA/04/13, p 8

Ray R, Ganguly D, Chowdhury C, Dey M, Das S, Dutta MK et al (2011) Carbon sequestration and annual increase of carbon stock in a mangrove forest. Atmos Environ 45(28):5016–5024. https://doi.org/10.1016/j.atmosenv.2011.04.074

Ren H, Wu X, Ning T, Huang G, Wang J, Jian S, Lu H (2011) Wetland changes and mangrove restoration planning in Shenzhen Bay, Southern China. Landsc Ecol Eng 7(2):241–250. https://doi.org/10.1007/s11355-010-0126-z

Short FT, Wyllie-Echeverria S (1996) Natural and human-induced disturbance of seagrasses. Environ Conserv 23:17–27. https://doi.org/10.1017/S0376892900038212

Thivakaran GA, Sawale AA, Asari RV (2016) Mangrove manual for gulf of Kachchh- a handbook for mangrove planters. Gujarat Institute of Desert Ecology, Bhuj, Gujarat. ISBN: 978-93-5258-954-8

Thivakaran GA, Sharma SB, Chowdhury A, Murugan A (2018) Status, structure and environmental variations in semi-arid mangroves of India. J For Res:1–11. https://doi.org/10.1007/s11676-018-0793-4

Tomlinson PB (2016) The botany of mangroves. Cambridge University Press. https://doi.org/10.1017/CBO9781139946575

Turner WR, Oppenheimer M, Wilcove DS (2009) A force to fight global warming. Nature 462 (7271):278. https://doi.org/10.1038/462278a

Valiela I, Bowen JL, York JK (2001) Mangrove forests: one of the world's threatened major tropical environments. Bioscience 51:807. https://doi.org/10.1641/0006-3568(2001)051[0807,MFOOTW]2.0.CO;2

Van der Werf GR, Morton DC, DeFries RS, Olivier JG, Kasibhatla PS, Jackson RB et al (2009) CO_2 emissions from forest loss. Nat Geosci 2(11):737. https://doi.org/10.1038/ngeo671

Van Nedervelde F, Cannicci S, Koedam N, Bosire J, Dahdouh-Guebas F (2015) What regulates crab predation on mangrove propagules? Acta Oecol 63:63–70

Walters BB (2000) Local mangrove planting in the Philippines: are fisherfolk and fishpond owners effective restorationists? Restor Ecol 8(3):237–246. https://doi.org/10.1046/j.1526-100x.2000.80035.x

Waycott M, Duarte CM, Carruthers TJB, Orth RJ, Dennison WC et al (2009) Accelerating loss of seagrasses across the globe threatens coastal ecosystems. Proc Natl Acad Sci U S A 106:12377–12381

Chapter 8
Biowaste for Carbon Sequestration

Nhamo Chaukura

Abstract The release of carbon into the environment contributes to greenhouse gases, which cause environmental problems such as climate change. Sources of carbon into the environment include the decomposition and burning of carbon-based materials such as biomaterials, fossil fuels, and municipal waste. Industrial operations such as coal powered power stations, breweries, pulp and paper, and tanning consequently have a high carbon footprint. The process of carbon sequestration captures and stores carbon to prevent it from entering the atmosphere. The sequestered carbon is transferred to a stable carbon pool to reduce carbon emission. One key greenhouse gas is CO_2, which causes global warming and climate change. In addition to CO_2 capture at point sources, its capture from the atmosphere may be necessary because approximately 50% of CO_2 emissions emanate from diffuse sources. Besides, CO_2 mitigation efforts have not progressed fast enough because of the wide usage of fossil fuels. In addition, CO_2 has a long residence time in the atmosphere so that climate change will persist even after total abatement of CO_2 emissions. Methods of mitigating carbon emissions include limiting the emission of carbon into the environment, and removing carbon from the environment. Owing to its high carbon content, biomass is a store of carbon, and can be used to sequester carbon from the environment. Carbon sequestration methods using biomass include composting, recycling, making biochar, and generating biofuels. A considerable number of anthropogenic activities such as food processing industry, pulp and paper industry, petroleum industry, and dyeing industry produce biomass as a waste. These biowastes can be used in the same way as pristine biomass to sequester carbon from the environment. In this chapter, the use of biowastes in the sequestration of carbon from environmental compartments is evaluated. The aims were to: (1) provide a brief overview of sources of biowastes, (2) assess the environmental impact of biowastes, and (3) evaluate the use of biowastes for carbon sequestration.

N. Chaukura (✉)
Materials Research (MatRes) Group, Chemistry Department, Bindura University of Science Education, Bindura, Zimbabwe

Nanotechnology and Water Sustainability (NanoWS) Research Unit, University of South Africa (UNISA), Johannesburg, South Africa
e-mail: nchaukura@buse.ac.zw

Inamuddin et al. (eds.), *Sustainable Agriculture Reviews 37*, Sustainable Agriculture Reviews 37, https://doi.org/10.1007/978-3-030-29298-0_8

Keywords Biochar · Bioenergy · Circular bioeconomy · Climate change · Composting · Environmental impacts · Global warming · Greenhouse gases · Life-cycle analysis · Mitigation

8.1 Introduction

The problem of climate change is growing at a concerning rate, increasing the depth of the ocean, altering ecosystems, and increasing the danger associated with storms (Kahiluoto et al. 2011; Mohan et al. 2018). This is caused by greenhouse gases, which are mostly of carbon origin. Greenhouse gases accumulate in the atmosphere where they trap heat from the sun, preventing it from escaping into space. Gases such as CO and CO_2 are released into the atmosphere from the combustion of fossil fuels such as coal, diesel, and petrol. Thus climate change is associated with the emission of air pollutants from human activities. Generally, biological treatment of organic wastes, and agricultural activities are key sources of CO_2, CH_4, and N_2O emissions (Thangarajan et al. 2013; Jensen et al. 2017). The soil is a repository of the largest pool of bio-available carbon in the form of humus. Conventional agricultural practices such as crop harvesting, excessive application of fertilizer, and soil tillage remove considerable amounts of humus from the soil through accelerated decomposition and soil degradation (Mandal et al. 2016). The carbon thus released enters the atmosphere where it contributes to climate change. Efficient biomass recycling, and the consequent reduction in fertilizer use result in a reduction in nutrient surplus, increased crop production, water retention, and the biodiversity of soil decomposers (Kahiluoto et al. 2011).

In the United States, the biggest source of greenhouse gas emissions is the transportation sector, followed by power generation, the manufacturing industry, and buildings (Rhodium Group 2018). An average car emits about 167.2 g CO_2 equ/km, and about 3344 kgCO_2 equ/year (Mühle et al. 2010). In China, CO_2 emissions rose by 4.52 times in the period 1997–2012 (Li et al. 2018), while energy-related CO_2 emissions in the United States increased by an estimated 3.4% in 2018, the largest rise in 8 years (Rhodium Group 2018). In 2004, per capita CO_2 emissions in the Republic of Chad were 0.0127 tCO_2 relative to an average of 1.0215 tCO_2 in sub-Sahara Africa (Couth and Trois 2010). Overall, the global concentration of CO_2 in the atmosphere has risen by more than 40% from around 1840, and is rising at approximately 2 ppm/year (Marmo 2008). Currently, CO_2 emissions are around 30 Gt/year globally, thus creating an imbalance in the carbon cycle resulting in a buildup in the atmosphere. In 2007, 28.8 Gt was emitted, and is projected to rise to 40.3 Gt by 2030, and farther to 50 Gt by 2050 if appropriate mitigation measures are not taken (Mohan et al. 2018). Policies were formulated to limit the increase of global temperature to 2 °C above pre-industrial levels based primarily on the reduction of greenhouse gas emissions. Furthermore, the European Union committed to achieve 20% decrease in greenhouse gas emissions (relative to 1990) up to 2020 (Moreira and Pires 2016).

In order to generate enough carbon currency to sustain the rapidly growing human population, fossil fuel carbon has been mechanically translocated from underground to readily accessible pools. However, tempering with the carbon cycle results in the greenhouse effect, with established environmental impacts (Moreira and Pires 2016). Although activities such as car-pooling and buying locally grown foods minimise carbon emissions into the atmosphere, it is also important to remove carbon from the atmosphere (carbon sequestration). Carbon sequestration can be achieved by using either: (1) CO_2-grabbing chemicals dissolved in water; or (2) high surface area solid support materials with CO_2-grabbing chemicals. The scale of carbon capture is so vast the materials have to be regenerated so that the carbon capture process is economically feasible and environmentally sustainable. Recycling carbon-capture materials consumes energy, and produces highly pure CO_2 which liquefies easily, is easier to handle and to use as a fuel or chemical feedstock. Interestingly, CO_2 is a potential precursor material in the bio-economy, which comprises the generation of renewable biogenic resources and their transformation into value-added products such as chemical feedstocks, bio-based products, bioenergy, and food (Sterner and Fritsche 2011; Mohan et al. 2018). Thus the sequestration of CO_2 can result in the production of a range of materials such as biomass, bio-fuels, polymers, and other chemicals (Rana et al. 2016; Mohan et al. 2018). The thermal energy required to recycle carbon capture materials determines the overall cost of the technology.

One carbon sequestration method uses biomass. Plants convert carbon from the atmosphere into biomass, which can be turned into charcoal by controlled combustion (Thangarajan et al. 2013). Carbon as charcoal is stable, and is not easily released into the atmosphere. Another carbon sequestration method involves capturing it before it is released into the atmosphere and storing it underground. There are many other sequestration methods, which all remove carbon from the atmosphere and transfer it to a passive pool that is inert and less harmful to the environment (Qambrani et al. 2017).

Biomass exists in organic matter from plants, and it comprises agricultural and forestry residues, energy crops, organic waste, and sewage sludge. It constitutes a promising renewable energy resource that can be used to produce fuels and electrical energy. Because it is a product of the reaction between CO_2, H_2O in the presence of sunlight via photosynthesis, biomass recycles the precursor materials, making CO_2 available again through respiration to produce new plants, thus reducing greenhouse gas emissions (Ramos et al. 2018). Overall, nature removes CO_2 from the atmosphere through sea water, soils, plants, and rocks. In order to cope with the increasing carbon emission, these natural processes have to be accelerated.

Because approximately 50% of the dry weight of microalgal biomass is carbon, microalgae and its derivatives can be used to mitigate CO_2 emissions (Moreira and Pires 2016). Some microalgae species can tolerate high concentrations of CO_2 in flue gases and have thus been incorporated into carbon capture technology. Their high carbon fixing efficiency and biomass productivity lends microalgae to applications in the reduction of greenhouse gas emission while simultaneously providing a renewable feedstock for biofuels and other materials (Yu et al. 2017; Mohan et al.

2018). Besides, agri-wastes and industrial by-product biomasses have potential for nutrient recycling and energy recovery, to mitigate eutrophication and climate change (Kahiluoto et al. 2011).

Biochar, a pyrogenic carbonaceous material derived from biomass and biowastes, can be used for carbon sequestration. Unlike the carbon in most organic matter, the chemistry of the carbon in biochar is modified by aromatization during pyrolysis to produce structures that are recalcitrant to microbial degradation. Consequently, biochar is stable for long periods of time, and should be effective for long-term carbon sequestration (Qambrani et al. 2017). The choice of feedstock for the production of biochar is influenced by the availability of biomass and biowastes, and the cost of gathering and transportation. Biochar feedstocks in sub-Saharan Africa include aquatic weeds, crop residues, livestock litter, forestry residues, agricultural processing wastes, municipality solid waste, and biosolids (Gwenzi et al. 2015).

In this chapter, the use of biowastes in the sequestration of carbon from environmental compartments is evaluated. The aims were to: (1) evaluate sources of biowastes, (2) assess the environmental impact of biowastes, and (3) evaluate the use of biowastes for carbon sequestration.

8.2 Sources of Biowastes

Owing to increasing global population and economic growth, large volumes of biowastes such as biosolids, livestock litter, agricultural processing wastes, municipal solid waste, and paper mill sludge, are currently being produced (Kahiluoto et al. 2011; Chaukura et al. 2017). Municipal waste is made up of two main sources of biogenic wastes, which are municipal solid waste and biosolids. Only part of the municipal solid waste is recycled, while the rest is discharges into the environment. Globally, the production of municipal solid waste increased from 1.3 billion tons in 1994 to 1.7 billion tons in 2008 (Mandal et al. 2016). On average, organic matter for urban municipal solid waste in Africa constitutes about 56%, and its decomposition is a significant source of greenhouse gases (Couth and Trois 2010). Converting biological waste into new materials addresses both environmental and sustainability concerns. For instance, municipal solid waste comprises >50% biodegradable material, thus it can be used as a feedstock for producing bioenergy, biofuel, and various chemicals. Likewise, the properties of biosolids lend them to applications as feedstock in the circular bioeconomy framework (Mohan et al. 2018).

The flow of material in the agrifood system are a significant element of the biodegradable material flows of modern civilisation. The agricultural food production and processing residues represent an ill-exploited resource. Compared to forest residues, their use in energy production is insignificant, and nutrient recycling is lacking. For instance, although nearly all livestock litter is returned to the soil, it is applied excessively to small sections of land so that the nutrient up-take by crops is low and it is necessary to supplement with inorganic fertilizers. As a result, there is nutrient loss into surface and groundwater. Agri-wastes and residues can replace non-renewable energy sources and fertilizers as demonstrated in biorefineries, which use the biomass in producing energy and various materials (Mohan et al. 2018). The potential for the biomass available in a particular region to mitigate climate change and nutrient loss sets the boundary conditions for the application of the technology, and is thus of primary interest. Technologies for greenhouse gas mitigation can be tailored to replace non-renewable energy or to allow considerable carbon sequestration and recycling (Kahiluoto et al. 2011).

8.3 Environmental Impact of Biowastes

Large amounts of municipal solid waste and biosolids are being produced and discharged into the environment daily. Waste dumps and landfills release greenhouse gases together with substantial volumes of leachate. Moreover, the burning of crop residues has been reported to cause smog, and cooking with firewood pollutes indoor air, causing health concerns in rural communities, especially in low income countries (Mohan et al. 2018).

The biological treatment of biowastes involves the release of greenhouse gases, which contribute to climate change. The extent of the greenhouse gas emissions impacts the environmental performance of the treatment process. Naturally, gas emissions from biological treatment processes are diffusive and usually take place via composting, leakages, material storage areas, and ventilated buildings (Prasad 2016; Jensen et al. 2017). The measurement of these emissions is limited by the diffusive and dynamic nature, and the large-scale occurrence of the emissions. Possible quantitation approaches include on-site measurement which determines emissions from different point sources at the facility, and remote sensing (Jensen et al. 2017).

The carbon footprint of a good or service is assessed from the energy expended during the lifecycle of material extraction, production, delivery and disposal. This can be computed basing on energy analysis, which uses either input-output analysis, process analysis, or an integrated life cycle assessment method, combining these two. This method is useful in estimating the carbon footprint indicator as a measure of the environmental impact of the production, disposal, and recycling of waste (Browne et al. 2009; Colon et al. 2010). For example, for an Irish city-region, the footprint of the disposal and recycling of waste decreased by about 7.7% in the period 1998–2004 (Browne et al. 2009). The carbon footprint approach is an element of ecological footprint analysis, which is a cumulative measure of the land footprint necessary to produce natural resources and sustenance infrastructure, and the land for greenhouse gas sequestration (Browne et al. 2009; Bergeron 2016).

Recently, the issue of climate change has attracted significant research attention and political priority. Barring some disagreements, the popular belief is that the rise in CO_2 concentrations in the atmosphere witnessed from the time of the industrial revolution is primarily responsible for climate change (Marmo 2008). Therefore, it is crucial to formulate policies that reduce greenhouse gases emissions (reduction of sources) and remove greenhouse gases from the atmosphere (creation of sinks). Ultimately, the goal of the United Nations Framework Convention on Climate Change (UNFCCC) was to prevent hazardous human tampering with the climate system. In line with this objective, the Kyoto Protocol of the UNFCCC set up emission limits for developed countries, in which the EU was required to reach an 8% reduction in CO_2 emissions from 2008 to 2012 compared to 1990 (Marmo 2008). Although developing countries such as in Africa have ratified the United Nations treaty, the implementation of such is limited by the availability of funding and incentives, among other factors. Moreover, in 1999, the European Union Landfill Directive made a requirement to reduce the quantity of disposed biogenic waste, encouraging measures that promote sorting activities at source, such as material recovery and recycling (Council of the European Union 1999; Colon et al. 2010). In the same vein, the European legislation (Directive 2008/98/EC) classifies recycling/reclamation of non-solvent organic materials such as composting and other biological conversion processes as wastes recovery processes (Bernal et al. 2017). The current European Union waste policy encourages the prevention, re-use, recycling, and recovery of waste. Priority is given to integrated material and energy recovery, and landfilling is a final alternative (Directive 2008/98/EC, article 4 2008). Should specific waste streams be diverted from this hierarchy, such justification should be based on life cycle and environmental impact assessment (Jensen et al. 2017). Other country-specific initiatives have been established. For example, in Poland, the Lodz Declaration introduced the concept of sustainable and renewable energy systems called Circular Bioregions, which aim at attaining a net zero energy balance with concurrent low CO_2 emission (Bis et al. 2018). Incentives such as tax rebates and financial subsidies can facilitate the achievement of carbon-neutral technologies. For instance, on account of the financial subsidies emanating from the German law on renewable energy, biodegradable waste from the material recycling chain is fed back into the energy recovery loop in biomass power stations (Kranert et al. 2010). Alternatively, imposing a carbon tax potentially raises the cost associated with carbon emissions and increases the demand and affordability of carbon-neutral processes. Nonetheless, this is only lucrative if the reduction costs do not exceed the tax (Moreira and Pires 2016).

8.4 Application of Biowastes for Carbon Sequestration

8.4.1 Composting Technology

Despite the landfill directive of 1999 in the European Union, about 40% of biomass waste is disposed in landfills (Eurostat 2010). Another treatment approach for biogenic household waste is composting, where biowaste is extracted from the waste stream at source, subsequently reducing the quantity of municipal waste

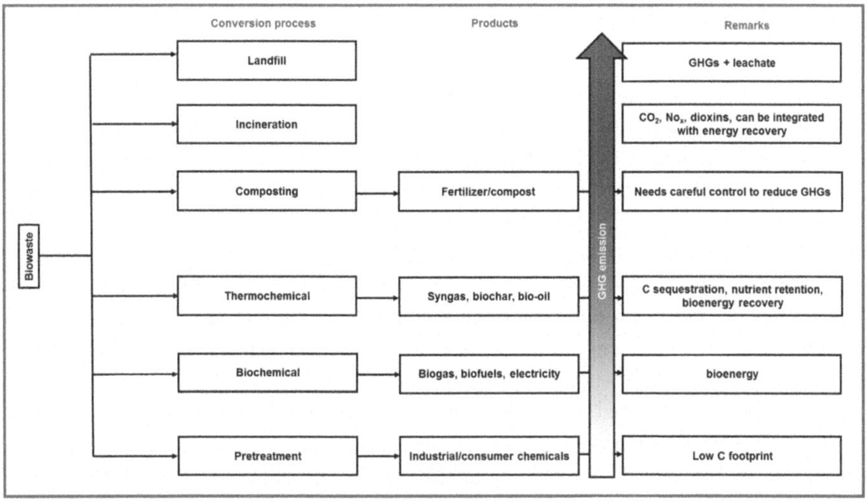

Fig. 8.1 Biowaste disposal and valorization processes

(Fig. 8.1) (Quirós et al. 2014). While composting of the organic fraction of municipal solid waste has been up-scaled to industrial scale (Kranert et al. 2010), the technical and scientific study of composting at household level is still lacking. The benefits of home composting include: (1) it avoids the collection of a key component of municipal solid waste, thus decreasing the economic, material and energy requirements in infrastructure. This suggests reduced land use and facilitates a more exact control of the composting process and the organic wastes feedstock (Colon et al. 2010). Thus home composting has a low carbon footprint. Generally, composting waste biomass stabilizes organic matter, reduces moisture content, increases the nutrient concentrations, eliminates pathogenic organisms and weed seeds, develop disease resistance, and reduces greenhouse gas emissions (Bernal et al. 2017). Because of the rich nutrient content, compost can be useful as a fertilizer in organic farming, thereby reducing the use of fossil-based fertilizers (Thangarajan et al. 2013; Paetsch et al. 2016; Mohan et al. 2018). However, if not well managed, composting can produce excessive gaseous emissions of NH_3, N_2O, and CH_4 (Malinska et al. 2014; Quirós et al. 2014; Bernal et al. 2017). Thus gaseous emissions are a critical parameter influencing the environmental performance of horticultural systems (Quirós et al. 2014). The quality of the resulting compost depends on the control of the composting process and environmental conditions. In order to minimize the risk of pollution due to composting and the application of compost, protocols such as the European Commission Communication on Biowaste, that provide quality procedures, have been developed (Bernal et al. 2017). The compost feedstock, mixing frequency, moisture content and temperature are the major variables that affect the quantity of emissions (Quirós et al. 2014).

8.4.2 As a Fertilizer/Organic Farming

While conventional agricultural practices can lead to land degradation and soil loss, organic farming is attractive because of its potential environmental benefits. Organic farming reduces soil loss, and increases soil organic matter and water retention, improves the soil microbial ecology, is more energy efficient, and generates less N_2O (Thangarajan et al. 2013; Carlson et al. 2017).

Perhaps the main disadvantage of organic farming is its low yields as a result of lower input levels, and increased pests and weeds. Consequently, organic farming uses larger areas of land to produce the same amount of output compared to conventional farming. It is therefore, important to implement practices capable of providing environmental benefits while sustaining high outputs (Tuomisto et al. 2012).

Anaerobically digested food waste is an alternative fertilizer (Fig. 8.1) that is rapidly gaining acceptance for both conventional and organic farming practices, especially in Sweden (Chiew et al. 2015). In comparison to mineral fertilizers, fertilizer derived from anaerobically digested food waste has a number of environmental benefits, as high quality energy is gained in the production process and the nutrients are maintained within the digestate. Although organic fertilization, particularly biogas digestate, has potential to increase N_2O and CO_2 emissions relative to mineral fertilizers, the production of synthetic fertilizers has a high energy demand, resulting in significant indirect greenhouse gas emissions (Chiew et al. 2015; Heintze et al. 2017).

8.4.3 Energy

Globally, about 80% of the total energy supply is from fossil fuels with concomitant high CO_2 emissions (Ramos et al. 2018). Fossil fuel scarcity with the associated climate change concerns have motivated the development of renewable energy technologies worldwide (Tong et al. 2018a, b).

In 2008, the Organisation for Economic Co-operation and Development reported that, the major countries producing electricity from biomass are in the order: United States (26%) > Germany (15%) > Brazil ~Japan (7%) (Ramos et al. 2018). Overall, there is a gradual shift from a linear fossil-based to a circular bio-based economy (Mohan et al. 2018).

One way to reduce the volumes of greenhouse gas emissions is through using renewable energy sources instead of fossil fuels in energy generation (Fig. 8.1) (Pertl et al. 2010; de Caprariis et al. 2017; Onorevoli et al. 2018). In this regard, the European Union approved the "Renewable Energy Directive" in 2008 as part of a broader climate change and energy strategy. This aimed to address the renewable energy industry's contribution towards attaining the Commission goal of the generation of 20% of energy from renewable sources by 2020 (Pertl et al. 2010). Biomass is a renewable energy source with great promise, which can be sustainably used in producing fuels and electrical energy with no carbon emissions.

A number of different ways such as oxidation through combustion, biochemical conversion (e.g., fermentation to obtain biogas or bio-hydrogen), and thermo-chemical conversion (e.g., gasification to produce syngas) can be used to convert biomass to bioenergy (Fig. 8.1) (Sterner and Fritsche 2011; Bogusz et al. 2017; Mohan et al. 2018; Ramos et al. 2018). Gasification is a thermal technique that converts biomass residues to valuable products, and thus helps prevent the depletion of non-renewable sources (Ramos et al. 2018). Moreover, technologies that convert waste to energy also reduce environmental pollution through decreasing the quantities of biogenic waste streams going to landfills and increasing reuse and recycling of waste (Shen et al. 2015). Landfills are a significant anthropogenic source of CH_4 emission in the, constituting about 18% of the CH_4 emissions in 2013 (United States Environmental Protection Agency 2015). Recycling and valorizing waste helps reduce the number of landfill sites, so much so that CH_4 emissions can be reduced as a result. Energy production from such processes can counterbalance the greenhouse gas emissions from waste treatment systems, resulting in a carbon-negative economy (Shen et al. 2015).

Biomass conversion into bioenergy integrated with carbon capture and storage has potential for CO_2 mitigation and keeping CO_2 in geological reservoir sinks. Owing to their high photosynthetic efficiencies, algae have high growth rates and high biomass yields, and as such, can be used as a biomass source with a broad range of applications. This enhances the economic feasibility of the conversion process (Moreira and Pires 2016). Another abundant source of biomass is sewage sludge. Owing to the environmental challenges associated with conventional sludge disposal methods such as land application and landfilling, the conversion of sewage sludge to energy represents a renewable energy source since it is produced in large quantities and has a significant energy content. Energy from sewage sludge is considered carbon-neutral and eco-friendly (Tong et al. 2018a). Landfilling produces fugitive emissions such as leachate and landfill gas, and potentially contaminates the environment, while land application is a major route for the input of pollutants such as toxic metals and a gamut of organic pollutants (Cao and Pawłowski 2013).

The availing of financial subsidies for the production of energy from renewable feedstocks imply that biogenic waste will be diverted in increasing amounts from material recovery and introduced into energy recovery systems through ignition at biomass power stations, for instance (Kranert et al. 2010; Cong et al. 2017). Singapore, for example has continuously made in-roads in increasing renewable energy generation to reduce expenditure on importing fossil fuels. While they might not replace natural gas power plants, they can help improve energy sufficiency and environmental sustainability. Governments in countries such as Zimbabwe and India have made the blending of bioethanol and gasoline mandatory, resulting in an increase in bioethanol plants (Mohan et al. 2018). The CO_2 emission accompanying biomass use is not regarded as a contributor to global warming, since the CO_2 released during biomass valorization is offset by the CO_2 withdrawn from the atmosphere during photosynthesis (Heintze et al. 2017; Tong et al. 2018a).

8.4.4 Biochar Technology

Biochar (Fig. 8.1) is a pyrogenic carbonaceous material with a high specific surface area and porous structure (Li et al. 2018; Rajapaksha et al. 2019; Restuccia et al. 2019; Yang et al. 2019). Its applications include as a soil amendment that offers carbon sequestration capabilities, and could potentially mitigate N_2O emissions (Dijkstra et al. 2012; Mandal et al. 2018; Speratti et al. 2018; Zhou et al. 2018; Briens and Bowden-Green 2019; Lan et al. 2019). Biochar application in soils is a simple carbon sequestration method that also enhances the soil's physico-chemical characteristics for improved plant growth (Bis et al. 2018; Sadeghi et al. 2018). After being produced through photosynthesis, efficient processing of biomass recycles the original compounds, leaving CO_2 available to produce new biomass (Qambrani et al. 2017; Ramos et al. 2018). The carbon in biochar is an inert carbon pool, with average residence times covering centuries to millennia (Dijkstra et al. 2012). Controlled pyrolysis is used for valorizing biomass such as sewage sludge, agri-wastes, forest residues, energy crops to useful products such as energy, bio-oil, syngas, and biochar. The relative yields and properties of the resulting materials are influenced by pyrolysis conditions such as temperature, heating rate and feeding mode, and on the characteristics of the feedstock. Biochar can be applied in carbon sequestration since it retains a significant amount of stable carbon (Cao and Pawłowski 2013). Overall, biochar technology has several strengths including: (1) it is a key component of circular bioregions' economy, (2) it is a powerful technique for clean energy conversion, and (3) it is an atmospheric carbon sink that mitigates climate change (Bis et al. 2018).

8.5 Future Research Directions

Techniques such as life cycle assessment are usually used to compare environmental impacts of contrasting systems. Life cycle assessment uses a "cradle-to-grave" approach in simultaneously accounting for a number of environmental features of a product or service (Rana et al. 2016). However, when alternative options are not properly considered the interpretation of these studies may be distorted (Tuomisto et al. 2012). Results of a life cycle assessment study have high uncertainty owing to the cumulative effects of data variation, erroneous measurements, incorrect estimations, missing or unrepresentative data, and intrinsic assumptions of models (Clavreul et al. 2012). Developments in uncertainty analysis in life cycle assessment studies have made significant progress over the last years, and in waste management, various approaches have been used but a more work to develop a systematic method of waste-life cycle assessment studies needs to be done (Clavreul et al. 2012). Future research should therefore focus on eliminating uncertainty in life cycle assessment in order to improve the accuracy of the prediction of environmental impacts and their subsequent alleviation.

Simulation and modelling approaches can be used to estimate the dynamics that affect greenhouse gas emissions at a facility such as a farm or manufacturing operations. A cradle-to-gate carbon calculator, such as OFoot, that quantifies greenhouse gas emissions of the consumables, equipment, and infrastructure used for farm operations; including the embodied energy and greenhouse gas emissions for each item, has been used (Carlson et al. 2017). Moreover, based on user-specified management practices, OFoot can be used to evaluate soil and crop dynamics and attendant greenhouse gas emissions (Carlson et al. 2017). Application of such modelling approaches can predict emissions associated with particular operations and practices and provide data can be used to reduce them. Further studies should develop cradle-to-gate carbon calculation methods that are user-friendly, and have capabilities to handle multiple variables that influence greenhouse gas emissions at a range of facilities including waste treatment operations, manufacturing industries, and service industries.

An important facet of a successful bio-economy is a sustainable business model interlinked with government policies, subsidies and penalties at the early stages to commercialize the technologies. In the case of solid waste, local governments are the main stakeholder who should look for revenue generation opportunities through public-private schemes. For example, in Sweden, business models of municipally-owned municipal solid waste operations integrate three types of activities: public service activities for collecting household and industrial solid waste; processing activities that convert the waste into valuable materials and products; and marketing and sales activities that facilitate the re-entrance of the products and recycled materials into the economy. An innovative approach would be establishing a business model based on mini-grids that are fed by bioenergy for rural communities with no access to the grid but located within a reasonable distance from biomass sources such as agricultural or forest residues (Sterner and Fritsche 2011; Mohan et al. 2018). This approach would be suitable especially for low income countries, where a consideration proportion of the population lives off the grid. The bioeconomy in India is expected at an annual growth rate of about 20% to over $100 billion by 2025 (Mohan et al. 2018). The main drivers for the Indian bioeconomy include high population growth rate, increasing consumerism, growing demand for energy and various products, and the rising generation of waste (Mohan et al. 2018). This trend is likely to be reproduced across the globe. Future research should establish the key parameters to be considered in a business model for bioenergy generation and other alternative uses of biowastes, especially in low income countries.

The deployment of bioenergy with carbon capture and storage technology depends on government regulations and policies, and public buy-in. The impacts of climate change have been increasing, resulting in a gradual shift in the attitudes of the public. Nonetheless, the rapid development of mitigation technologies also depends on the economic instruments, which are normally profit driven. Therefore, uncertainties such as the complexity, wide geographic coverage and a long-term presence, discourage private investment. Similarly, governments are cautious, withholding investment even more. At the onset, these technologies are not economically competitive, but with financial support and public uptake, they should pick up and start generating profits (Moreira and Pires 2016). Going forward, research should focus on the formulation of government policies on waste disposal and valorization; how they can facilitate the economic feasibility and public acceptance of carbon capture technologies.

8.6 Conclusion

Instead of treating biowaste as a problem, it should be considered as an opportunity. Rather than costly disposal which can cause adverse environmental effects, waste biomaterial can be valorized through a number of processing approaches. Biogenic waste is progressively being diverted from the material recycling loop and, as a result of the financial subsidy arising from laws and policies in developed countries; it is fed back into the energy recovery processes such as biomass power stations. Besides, biorefineries for the production of recycled higher value products and bioenergy are gaining significance in developed countries. Landfilling has a number of issues associated with it: (1) land resources are under enormous pressure globally; (2) landfills generate greenhouse gases; and (3) resources that would otherwise be useful are being disposed. For a long time, waste management involved dumping, burning, composting, and converting waste into recycled products. Yet, valorizing biowaste to produce higher-value materials has recently attained great importance. Incineration and composting are no longer universally applicable.

Because it incorporates waste into a resource-based economy suitable to all types of waste, the circular economic model is proper for waste management. The main drivers in attaining these objectives include: (1) lower carbon footprint and reduced waste load, (2) cost of feedstock, carbon tax, or waste taxation, and (3) legal framework, branding, renewable. Overall, there is a gradual shift from a linear fossil-based to a ‘circular bio-based’ economy, which can be used for carbon capture and storage, thus assisting in greenhouse gas mitigation.

Future research should focus on: (1) improving the accuracy of the prediction of environmental impacts and their subsequent alleviation by eliminating uncertainty in life cycle assessment approaches, (2) developing cradle-to-gate carbon calculation methods that are user-friendly, and have capabilities to handle multiple variables that influence greenhouse gas emissions at a range of facilities, (3) establishing key variables to be considered in a business model for bioenergy generation and other alternative uses of biowastes, especially for low income countries, and (4) how government policies on waste disposal and valorization can facilitate the economic feasibility and public acceptance of carbon capture technologies.

References

Bergeron FC (2016) Energy and climate impact assessment of waste wood recovery in Switzerland. Biomass Bioenergy 94:245–257

Bernal MP, Sommer SG, Chadwick D, Qing C, Guoxue L, Michel FC (2017) Current approaches and future trends in compost quality criteria for agronomic, environmental, and human health benefits. In: Sparks DL (ed) Advances in Agronomy, vol 144. Academic, Cambridge, MA, pp 143–233

Bis Z, Kobyłecki R, Ścisłowska M, Zarzycki R (2018) Biochar – potential tool to combat climate change and drought. Ecohydrol Hydrobiol 18:441–453

Bogusz A, Nowak K, Stefaniuk M, Dobrowolski R, Oleszczuk P (2017) Synthesis of biochar from residues after biogas production with respect to cadmium and nickel removal from wastewater. J Environ Manage 201:268–276

Briens L, Bowden-Green B (2019) A comparison of drum granulation of biochars. Powder Technol 343:723–732

Browne D, O'Regan B, Moles R (2009) Use of carbon footprinting to explore alternative household waste policy scenarios in an Irish city-region. Resour Conserv Recycl 54:113–122

Cao Y, Pawłowski A (2013) A life cycle assessment of two emerging sewage sludge-to-energy systems: evaluating energy and greenhouse gas emissions implications. Bioresour Technol 127:81–91

Carlson BR, Carpenter-Boggs LA, Higgins SS, Nelson R, Stöckle CO, Weddell J (2017) Development of a web application for estimating carbon footprints of organic farms. Comput Electron Agric 142:211–223

Chaukura N, Murimba EC, Gwenzi W (2017) Synthesis, characterisation and methyl orange adsorption capacity of ferric oxide-biochar nano-composites derived from pulp and paper sludge. Appl Water Sci 7:2175. https://doi.org/10.1007/s13201-016-0392-5

Chiew YL, Spångberg J, Baky A, Hansson PA, Jönsson H (2015) Environmental impact of recycling digested food waste as a fertilizer in agriculture – a case study. Resour Conserv Recycl 95:1–14

Clavreul J, Guyonnet D, Christensen TH (2012) Quantifying uncertainty in LCA-modelling of waste management systems. Waste Manag 32:2482–2495

Colón J, Martínez-Blanco J, Gabarrell X, Artola A, Sánchez A, Rieradevall J, Font X (2010) Environmental assessment of home composting. Resour Conserv Recycl 54:893–904

COM(2010)235. https://www.eumonitor.eu/9353000/1/j9vvik7m1c3gyxp/vikqhoczvsub. Accessed 20 Feb 2019

Cong R, Caro D, Thomsen M (2017) Is it beneficial to use biogas in the Danish transport sector? - an environmental-economic analysis. J Clean Prod 165:1025–1035

Council of the European Union (1999) Directive 1999/31/EC, of 26 April 1999 on the landfill of waste

Couth R, Trois C (2010) Carbon emissions reduction strategies in Africa from improved waste management: a review. Waste Manag 30:2336–2346

de Caprariis B, De Filippis P, Hernandez AD, Petrucci E, Petrullo A, Scarsella M, Turchi M (2017) Pyrolysis wastewater treatment by adsorption on biochars produced by poplar biomass. J Environ Manage 197:231–238

Dijkstra FA, Prior SA, Runion GB, Torbert HA, Tian H, Lu C, Venterea RT (2012) Effects of elevated carbon dioxide and increased temperature on methane and nitrous oxide fluxes: evidence from field experiments. Front Ecol Environ 10:520–527. https://doi.org/10.1890/120059

Directive 2008/98/EC of the European Parliament and of the Council of 19 November 2008 on waste and repealing certain Directives

Eurostat: Statistical Office of the European Communities (2010) Online database: http://epp.eurostat.ec.europa.eu/portal/page/portal/environmental/data/database. Accessed 14 Feb 2019

Gwenzi W, Chaukura N, Mukome FND, Machado S, Nyamasoka B (2015) Biochar production and applications in sub-Saharan Africa: opportunities, constraints, risks and uncertainties. J Environ Manage 150:250–261

Heintze G, Eickenssheidt T, Schmidhalter U, Drosler M (2017) Influence of soil organic carbon on greenhouse gas emission potential after application of biogas residues or cattle slurry: results from a pot experiment. Pedosphere 27:807–821. https://rhg.com/research/preliminary-us-emissions-estimates-for-2018/. Accessed 9 Jan 2019

Jensen MB, Møller J, Mønster J, Scheutz C (2017) Quantification of greenhouse gas emissions from a biological waste treatment facility. Waste Manag 67:375–384

Kahiluoto H, Kuisma M, Havukainen J, Luoranen M, Karttunen P, Lehtonen E, Horttanainen M (2011) Potential of agrifood wastes in mitigation of climate change and eutrophication – two case regions. Biomass Bioenergy 35:1983–1994

Kranert M, Gottschall R, Bruns C, Hafner G (2010) Energy or compost from green waste? – a CO_2-based assessment. Waste Manag 30:697–701

Lan ZM, Chen CR, Rashti MR, Yang H, Zhang DK (2019) Linking feedstock and application rate of biochars to N_2O emission in a sandy loam soil: potential mechanisms. Geoderma 337:880–892

Li S, Barreto V, Li R, Chen G, Hsieh YP (2018) Nitrogen retention of biochar derived from different feedstocks at variable pyrolysis temperatures. J Anal Appl Pyrolysis 133:136–146

Malińska K, Zabochnicka-Świątek M, Dach J (2014) Effects of biochar amendment on ammonia emission during composting of sewage sludge. Ecol Eng 71:474–478

Mandal S, Kunhikrishnan A, Bolan NS, Wijesekara H, Naidu R (2016) Application of biochar produced from biowaste materials for environmental protection and sustainable agriculture production. In: Prasad MNV, Shih K (eds) Environmental materials and waste: resource recovery and pollution prevention. Academic Press, London, pp 73–89

Mandal S, Donner E, Vasileiadis S, Skinner W, Smith E, Lombi E (2018) The effect of biochar feedstock, pyrolysis temperature, and application rate on the reduction of ammonia volatilisation from biochar-amended soil. Sci Total Environ 627:942–950

Marmo L (2008) EU strategies and policies on soil and waste management to offset greenhouse gas emissions. Waste Manag 28:685–689

Mohan VS, Chiranjeevi P, Dahiya S, Kumar NA (2018) Waste derived bioeconomy in India: a perspective. N Biotechnol 40(Part A):60–69

Moreira D, Pires JCM (2016) Atmospheric CO_2 capture by algae: negative carbon dioxide emission path. Bioresour Technol 215:371–379

Mühle S, Balsam I, Cheeseman CR (2010) Comparison of carbon emissions associated with municipal solid waste management in Germany and the UK. Resour Conserv Recycl 54:793–801

Onorevoli B, Maciel GPS, Machado ME, Corbelini V, Caramão EB, Jacques RA (2018) Characterization of feedstock and biochar from energetic tobacco seed waste pyrolysis and potential application of biochar as an adsorbent. J Environ Chem Eng 6:1279–1287

Paetsch L, Mueller CW, Rumpel C, Houot S, Kögel-Knabner I (2016) Urban waste composts enhance OC and N stocks after long-term amendment but do not alter organic matter composition. Agric Ecosyst Environ 223:211–222

Pertl A, Mostbauer P, Obersteiner G (2010) Climate balance of biogas upgrading systems. Waste Manag 30:92–99

Prasad MNV (2016) Recovery of resources from biowaste for pollution prevention. In: Prasad MNV, Shih K (eds) Environmental Materials and Waste. Academic, London, pp 1–19

Qambrani NA, Rahman MM, Won S, Shim S, Ra C (2017) Biochar properties and eco-friendly applications for climate change mitigation, waste management, and wastewater treatment: a review. Renew Sustain Energy Rev 79:255–273

Quirós R, Villalba G, Muñoz P, Colón J, Font X, Gabarrell X (2014) Environmental assessment of two home composts with high and low gaseous emissions of the composting process. Resour Conserv Recycl 90:9–20

Rajapaksha AU, Ok YS, El-Naggar A, Kim H, Song F, Kang S, Tsang YF (2019) Dissolved organic matter characterization of biochars produced from different feedstock materials. J Environ Manage 233:393–399

Ramos A, Monteiro E, Silva V, Rouboa A (2018) Co-gasification and recent developments on waste-to-energy conversion: a review. Renew Sustain Energy Rev 81(P1):380–398

Rana R, Ingrao C, Lombardi M, Tricase C (2016) Greenhouse gas emissions of an agro-biogas energy system: estimation under the renewable energy directive. Sci Total Environ 550:1182–1195

Restuccia F, Mašek O, Hadden RM, Rein G (2019) Quantifying self-heating ignition of biochar as a function of feedstock and the pyrolysis reactor temperature. Fuel 236:201–213

Rhodium Group (2018) Final US emissions estimates for 2018. https://rhg.com/research/final-us-emissions-estimates-for-2018/

Sadeghi SH, Panah MHG, Younesi H, Kheirfam H (2018) Ameliorating some quality properties of an erosion-prone soil using biochar produced from dairy wastewater sludge. Catena 171:193–198

Shen Y, Linville JL, Urgun-Demirtas M, Schoene RP, Snyder SW (2015) Producing pipeline-quality biomethane via anaerobic digestion of sludge amended with corn stover biochar with in-situ CO_2 removal. Appl Energy 158:300–309

Speratti AB, Johnson MS, Sousa HM, Dalmagro HJ, Couto EG (2018) Biochar feedstock and pyrolysis temperature effects on leachate: DOC characteristics and nitrate losses from a Brazilian Cerrado Arenosol mixed with agricultural waste biochars. J Environ Manage 211:256–268

Sterner M, Fritsche U (2011) Greenhouse gas balances and mitigation costs of 70 modern Germany-focused and 4 traditional biomass pathways including land-use change effects. Biomass Bioenergy 35:4797–4814

Thangarajan R, Bolan NS, Tian G, Naidu R, Kunhikrishnan A (2013) Role of organic amendment application on greenhouse gas emission from soil. Sci Total Environ 465:72–96

Tong H, Shen Y, Zhang J, Wang CH, Ge TS, Tong YW (2018a) A comparative life cycle assessment on four waste-to-energy scenarios for food waste generated in eateries. Appl Energy 225:1143–1157

Tong H, Yao Z, Lim JW, Mao L, Zhang J, Ge TS, Peng YH, Wang CH, Tong YW (2018b) Harvest green energy through energy recovery from waste: a technology review and an assessment of Singapore. Renew Sustain Energy Rev 98:163–178

Tuomisto HL, Hodge ID, Riordan P, Macdonald DW (2012) Does organic farming reduce environmental impacts? – a meta-analysis of European research. J Environ Manage 112:309–320

US Environmental Protection Agency (2015) Inventory of U.S. Greenhouse Gas Emissions and Sinks: 1990–2013. US Environmental Protection Agency, Washington, DC

Yang X, Ng W, Wong BSE, Baeg GH, Wang CH, Ok YS (2019) Characterization and ecotoxicological investigation of biochar produced via slow pyrolysis: effect of feedstock composition and pyrolysis conditions. J Hazard Mater 365:178–185

Yu KL, Show PL, Ong HC, Ling TC, Lan JCW, Chen WH, Chang JS (2017) Microalgae from wastewater treatment to biochar – feedstock preparation and conversion technologies. Energ Conver Manage 150:1–13

Zhou X, Liang C, Jia L, Feng L, Wang R, Wu H (2018) An innovative biochar-amended substrate vertical flow constructed wetland for low C/N wastewater treatment: impact of influent strengths. Bioresour Technol 247:844–850

Index

Inamuddin et al. (eds.), *Sustainable Agriculture Reviews 37*, Sustainable Agriculture Reviews 37, https://doi.org/10.1007/978-3-030-29298-0

Y

Z

The manufacturer's authorised representative in the EU is Springer Nature Customer Service Centre GmbH, Europaplatz 3, 69115 Heidelberg, Germany. If you have any concerns regarding our products, please contact ProductSafety@springernature.com

Printed and bound by CPI Group (UK) Ltd, Croydon, CR0 4YY
30/09/2026
02225608-0015